Distributed Parameter Modeling and Boundary Control of Flexible Manipulators

Jinkun Liu · Wei He

Distributed Parameter Modeling and Boundary Control of Flexible Manipulators

清華大学出版社
TSINGHUA UNIVERSITY PRESS

② Springer

Jinkun Liu
School of Automation Science and Electrical
 Engineering
Beihang University
Beijing
China

Wei He
School of Automation and Electric
 Engineering
University of Science and Technology
 Beijing
Beijing
China

ISBN 978-981-10-8299-3   ISBN 978-981-10-8300-6 (eBook)
https://doi.org/10.1007/978-981-10-8300-6

Jointly Published with Tsinghua University Press, Beijing, China

The print edition is not for sale in China Mainland. Customers from China Mainland please order the print book from: Tsinghua University Press.

Library of Congress Control Number: 2018934866

© Tsinghua University Press, Beijing and Springer Nature Singapore Pte Ltd. 2018
This work is subject to copyright. All rights are reserved by the Publishers, whether the whole or part of the material is concerned, specifically the rights of translation, reprinting, reuse of illustrations, recitation, broadcasting, reproduction on microfilms or in any other physical way, and transmission or information storage and retrieval, electronic adaptation, computer software, or by similar or dissimilar methodology now known or hereafter developed.
The use of general descriptive names, registered names, trademarks, service marks, etc. in this publication does not imply, even in the absence of a specific statement, that such names are exempt from the relevant protective laws and regulations and therefore free for general use.
The publishers, the authors and the editors are safe to assume that the advice and information in this book are believed to be true and accurate at the date of publication. Neither the publishers nor the authors or the editors give a warranty, express or implied, with respect to the material contained herein or for any errors or omissions that may have been made. The publishers remains neutral with regard to jurisdictional claims in published maps and institutional affiliations.

Printed on acid-free paper

This Springer imprint is published by the registered company Springer Nature Singapore Pte Ltd. part of Springer Nature
The registered company address is: 152 Beach Road, #21-01/04 Gateway East, Singapore 189721, Singapore

Preface

In recent years, distributed parameter system, boundary control techniques and their successful applications have a rapid development. Numerous theoretical studies and actual industrial implementations demonstrate that boundary control is a good candidate for control system design in solving the control problems of distributed parameter system. Many control approaches/methods, reporting inventions, and control applications within the fields of adaptive boundary control have been published in various journals and conference proceedings.

This book is motivated by the need for systematic design approaches for distributed parameter system using boundary control techniques. The main objectives of the book are to introduce the concrete design method and MATLAB simulation of boundary control strategies.

It is our goal to accomplish these objectives:

- Offer a catalog of implementable boundary control design methods for engineering applications.
- Provide advanced boundary controller design methods and their stability analysis methods.
- Offer simulation examples and MATLAB program for each boundary control algorithm.

This book provides the reader with a thorough grounding in the boundary controller design. Typical boundary controller design is verified using MATLAB simulation. In this book, concrete case studies, which present the results of boundary controller implementations, are used to illustrate the successful application of the theory.

The book is structured as follows. The book starts with a brief introduction of boundary control for mechanical systems in Chap. 1; mathematical preliminaries and design remarks are given in Chap. 2, in Chap. 3, PDE model and boundary conditions of flexible manipulator are given; in Chap. 4, a composite boundary controller for flexible manipulator is presented based on the PDE model. A boundary controller is designed with exponential convergence in Chap. 5, in Chap. 6, the boundary controller based on LaSalle analysis is designed, asymptotic stability of the closed-loop system can be guaranteed, in Chap. 7, full state constraints control is investigated for output constrained flexible manipulator system based on the PDE dynamic model. In Chap. 8, a boundary controller in the presence of control input constraint is designed to regulate angular position and suppress elastic vibration simultaneously. In Chap. 9, a robust observer based on the PDE dynamic model is given to estimate the distributed spatiotemporally varying states with unknown boundary disturbance and spatially distributed disturbance. In Chap. 10, an infinite dimensional disturbance observer based on the PDE dynamic model is introduced. In Chap. 11, a boundary controller design method with guaranteed transient performance is introduced, and the proposed control scheme allows the errors to converge to an arbitrarily small residual set, with convergence rate larger than a pre-specified value.

For each chapter, several engineering application examples are given. The contents of each chapter in this book are independent, so that readers can their own needs.

In this book, all the control algorithms and their programs are described separately and classified by the chapter name, which can be run successfully in MATLAB 7.5.0.342 version or in other more advanced versions. In addition, all the programs can be downloaded via http://shi.buaa.edu.cn/liujinkun. If you have questions about algorithms and simulation programs, please e-mail ljk@buaa.edu.cn.

This book was supported by the National Natural Science Foundation of China [grant number 61374048].

Beijing, China   Jinkun Liu
                                                                                                                          Wei He

Acknowledgements

First of all, we would like to express our gratitude to our co-workers who have contributed to the collaborative studies of this book. We would also like to express our sincere appreciation to our colleagues who have contributed to the collaborative research. In particular, we would like to thank Zhijie Liu and Fangfei Cao and Tingting Jiang, from Beihang University, China, for proofreading and providing numerous useful comments to improve the quality of the book.

Last but not least, we are deeply gratitude to our families for their invaluable loves, supports and sacrifices over the years.

This work is supported by

(i) "PDE modeling and adaptive boundary control theory research of N-link flexible manipulator," funded by the National Natural Science Foundation of China (NSFC) under the Grant 61374048, China.
(ii) "Design and Research on Boundary Control of a Flexible Marine Riser," funded by the National Natural Science Foundation of China (NSFC) under the Grant 61203057, China.

Beijing, China
August 2017

Jinkun Liu
Wei He

Contents

1 **Introduction** 1
  1.1   Control of Flexible Mechanical Systems 1
  1.2   Outline of the Book 3
  References 4

2 **Mathematical Preliminaries** 7
  2.1   The Hamilton Principle 7
  2.2   Functional and Variation 8
      2.2.1   Functional Variation Rules 8
      2.2.2   The Expansion of $\int_{t_1}^{t_2} \delta\left(\frac{1}{2}\dot{\theta}^2\right) dt$ 8
      2.2.3   Definition of Variation 9
  2.3   Discrete Simulation Method 9
      2.3.1   Discretization of Joint Angle $\theta(t)$ 10
      2.3.2   Several Discretization Methods 11
      2.3.3   Discretization of Boundary Conditions 11
  References 13

3 **PDE Modeling for Flexible Manipulator** 15
  3.1   PDE Modeling 15
  3.2   Simulation Example 21
  References 26

4 **Boundary Control for Flexible Manipulator Using Singular Perturbation** 27
  4.1   Introduction 27
  4.2   PDE Dynamic Model 27
  4.3   Singular Perturbed Dynamics 28
      4.3.1   Slow Subsystem 29
      4.3.2   Fast Subsystem 30

	4.4	Boundary Controller Design	30
		4.4.1 Controller for Slow Subsystem	31
		4.4.2 Controller for Fast Subsystem	32
		4.4.3 Total Boundary Controller	36
	4.5	Simulation Example	37
		References	43
5	**Boundary Control for Flexible Manipulator with Exponential Convergence**		45
	5.1	System Description	45
	5.2	Some Lemmas	46
	5.3	Boundary Controller Design	46
	5.4	Simulation Example	56
		References	63
6	**Boundary Control for Flexible Manipulator with LaSalle Analysis**		65
	6.1	System Description	65
	6.2	Dissipative Analysis of the Closed System	66
	6.3	Unique Analysis of Solutions	69
	6.4	Convergence Analysis	70
	6.5	Simulation Example	72
		References	77
7	**Boundary Control for Flexible Manipulator with State Constraints**		79
	7.1	Introduction	79
	7.2	System Statement	79
	7.3	Controller Design and Analysis	80
	7.4	Convergence Analysis	84
	7.5	Simulation Example	85
		References	93
8	**Boundary Control of Flexible Manipulator with Input Constraints**		95
	8.1	Introduction	95
	8.2	System Description	95
	8.3	Controller Design	96
	8.4	Dissipative Analysis of the Closed System	97
	8.5	Unique Analysis of Solutions	99
	8.6	Convergence Analysis	100
	8.7	Simulation Example	102
		References	108

9 Robust Observer Design for Flexible Manipulator Based on PDE Model ... 109
- 9.1 Introduction ... 109
- 9.2 System Description ... 109
- 9.3 Preliminaries ... 110
- 9.4 Observer Design and Analysis ... 111
- 9.5 Simulation Example ... 114
- References ... 123

10 Infinite Dimensional Disturbance Observer for Flexible Manipulator ... 125
- 10.1 Introduction ... 125
- 10.2 Observer Design ... 126
- 10.3 Simulation Example ... 128
- References ... 134

11 Boundary Control for Flexible Manipulator with Guaranteed Transient Performance ... 135
- 11.1 System Description ... 135
- 11.2 Preliminaries ... 136
- 11.3 Performance Function ... 137
- 11.4 Controller Design and Analysis ... 137
- 11.5 Convergence Analysis ... 141
- 11.6 Simulation Example ... 143
- References ... 152

12 Conclusions ... 153

About the Authors

Liu Jinkun was born on October 14, 1965, and he received BS, MS, and Ph.D. degrees from Northeastern University, Shenyang, China, in 1989, 1994, and 1997, respectively. He was a Postdoctoral Fellow in Zhejiang University from 1997 to 1999. He is currently a Full Professor in Beihang University, Beijing, P. R. China. His main research interest is adaptive boundary control for flexible manipulator. He has published more than 100 research papers and 8 books.

Professor Wei He received his PhD from Department of Electrical & Computer Engineering, the National University of Singapore (NUS), Singapore, in 2011, his M.Eng. and B.Eng. degrees both from School of Automation Science and Engineering, South China University of Technology (SCUT), Guangzhou, China, in 2008 and 2006 respectively. He is currently working as the Full Professor at School of Automation and Electric Engineering, University of Science and Technology Beijing (USTB), China. He is a senior member of IEEE. He has been awarded a Newton Advanced Fellowship from the Royal Society, UK in 2017. He is a recipient of the IEEE SMC Society Andrew P. Sage Best Transactions Paper Award in 2017. He is serving as the Associate Editor of IEEE Transactions on Neural Networks and Learning Systems, IEEE Transactions on Control Systems Technology, IEEE Transactions on Systems, Man, and Cybernetics: Systems, IEEE Access, the Editor of IEEE/CAA Journal of Automatica Sinica, Journal of Intelligent & Robotic Systems, Neurocomputing. He is the member of the IFAC TC on Distributed Parameter Systems, IFAC TC on Computational Intelligence in Control and IEEE CSS TC on Distributed Parameter Systems. His current research interests include robotics, distributed parameter systems and intelligent control systems.

Chapter 1
Introduction

1.1 Control of Flexible Mechanical Systems

In recent decades, dealing with the vibration problem of flexible systems has become an important research topic, driven by practical needs and theoretical challenges. Lightweight mechanical flexible systems possess many advantages over conventional rigid ones, such as lower cost, better energy efficiency, higher operation speed, and improved mobility. These advantages greatly motivate the applications of the mechanical flexible systems in industry. A large number of systems can be modeled as mechanical flexible systems such as telephone wires, conveyor belts, crane cables, helicopter blades, robotic arms, mooring lines, marine risers, and so on. However, unwanted vibrations due to the flexibility property and the time-varying disturbances restrict the utility of these flexible systems in different engineering applications.

Many physical processes cannot be modeled by ODEs since the state of the system depends on more than one independent variable. The state of a given physical system, such as flexible structure, fluid dynamics, and heat transfer, may depend on the time t and the location x. The flexible mechanical systems are dependent on the spatial and temporal variables, which can be modeled as the distributed parameter systems. The model is represented by a set of infinite-dimensional equations (i.e., PDEs describing the dynamics of the flexible bodies) coupled with a set of finite-dimensional equations (i.e., ODEs describing the boundary conditions). The model of the flexible mechanical system represented by a set of PDEs is difficult to control due to the infinite dimensionality of the system, and many control strategies for the conventional rigid-body system cannot be directly applied to solve the control problem of the flexible system.

The most popular control approaches for the distributed parameter systems are modal control based on the truncated discredited system model, distributed control by using distributed sensors and actuators, and boundary control.

Modal control for the distributed parameter systems is based on truncated finite-dimensional modes of the system, which are derived from finite element method, Galerkin's method, or assumed-modes method. For these finite-dimensional models, many control techniques developed for ODE systems can be applied. The truncated models are obtained via the model analysis or spatial discretization, in which the flexibility is represented by a finite number of modes by neglecting the higher-frequency modes. The problems arising from the truncation procedure in the modeling need to be carefully treated in practical applications. A potential drawback in the above control design approaches is that the control can cause the actual system to become unstable due to excitation of the unmodeled, high-frequency vibration modes (i.e., spillover effects) [1]. Spillover effects that result in instability of the system are investigated in [2, 3] when the control of the truncated system is restricted to a few critical modes. The control order needs to be increased with the number of flexible modes considered to achieve high accuracy of performance, and the control may also be difficult to implement from the engineering point of view since full state measurements or observers are often required.

In an attempt to overcome the above shortcomings of the truncated model-based modal control, boundary control where the actuation and sensing are applied only through the boundary of the system utilizes the distributed parameter model with PDEs to avoid control spillover instabilities. Boundary control combined with other control methodologies [4], such as variable structure control, sliding-mode control, energy-based robust control, model-free control, the averaging method, and robust adaptive control, is developed. In these approaches, system dynamics analysis and control design are carried out directly based on the PDEs of the system.

Distributed control [5] requires relatively more actuators and sensors, which makes the distributed control relatively difficult to implement. Compared with the distributed control, boundary control is an economical method to control the distributed parameter system without decomposing the system into a finite-dimensional space. Boundary control is considered to be more practical in a number of research fields, including the vibration control of flexible structures, fluid dynamics, and heat transfer, which requires few sensors and actuators. In addition, the kinetic energy, the potential energy, and the work done by the nonconservative forces in the process of modeling can be directly used to design the Lyapunov function of the closed-loop system.

In the literature of boundary control for the distributed parameter systems, functional analysis and the semigroup theory are usually used for the stability analysis and for the proof of the existence and uniqueness of PDEs, for example, in [6–8]. In [6], stability of different infinite-dimensional systems is studied based on the semigroup theory. In [7], stabilization of a second-order PDE system under noncollocated control and obser-vations is investigated in Hilbert spaces. With control at one end and noncollocated observation at the other end, the exponential stability of the closed-loop system is proved in [8].

Distributed parameter systems are described by operator equations on an infinite-dimensional Hilbert or Banach space [9, 10]. The stability analysis and the solution existence are based on the theory of semigroups on infinite-dimensional state spaces. In [11], the proof of the existence and uniqueness of the control system is carried out by using an infinite-dimensional state space. In [12], the asymptotic stability of the system with proposed control is proved by using semigroup theory. In [13], a noncollocated boundary control is developed to stabilize two connected strings with the joint anti-damping, and the exponentially stability is proved by using the semi-group theory.

Compared with the functional-analysis-based methods, the Lyapunov's direct method for the distributed parameter systems requires little background beyond calculus for users to understand the control design and the stability analysis. In addition, the Lyapunov's direct method provides a convenient technique for PDEs by using well-understood mathematical tools such as algebraic and integral inequalities and integration by parts.

The relevant applications for boundary control approaches in mechanical flexible structures consist of second-order structures (strings and cables) and fourth-order structures (beams and plates) [14]. The Lyapunov's direct method is widely used since the Lyapunov functionals for control design closely relate to kinetic, potential, and work energies of the distributed parameter systems. Based on the Lyapunov's direct method, the authors in [14–22] have presented results for the boundary control of the flexible mechanical systems. In [15], robust adaptive boundary control is investigated to reduce the vibration for a moving string with spatiotemporally varying tension. In [16], robust and adaptive boundary control is developed to stabilize the vibration of a stretched string on a moving transporter. In [17], a boundary controller for a linear gantry crane model with a flexible string-type cable is developed and experimentally implemented. An active boundary control system is introduced in [18] to damp undesirable vibrations in a cable. In [19], the asymptotic and exponential stability of an axially moving string is proved by using a linear and nonlinear state feedback. In [20], a flexible rotor with boundary control is illustrated, and the experimental implementation of the flexible rotor controller is also presented. Boundary control is applied to beams in [21], where boundary feedback is used to stabilize the wave equations and design active constrained layer damping. Active boundary control of an Euler–Bernoulli beam, which enables the generation of the desired boundary condition at any designators position of a beam structure, is investigated in [22]. In [23], a nonlinear control law is constructed to exponentially stabilize a free transversely vibrating beam via boundary control. In [24, 25], a boundary controller for the flexible marine riser with actuator dynamics is designed based on the Lyapunov's direct method and the backstepping technique.

1.2 Outline of the Book

The general objectives of the book are to develop constructive and systematic methods of designing control for flexible manipulators with guaranteed stability. By investigating the characteristics of several different models, control strategies are proposed to achieve the performance for the concerned systems. The book starts with a brief introduction of control techniques for flexible mechanical systems in this chapter. Chapter 2 presents several lemmas and properties for the subsequent development for the convenience of derives of the dynamical models and further stability analysis, discrete simulation methods also are introduced. In Chap. 3, PDE modeling for flexible manipulator is introduced. In Chap. 4, a composite boundary controller for flexible manipulator is presented based on the PDE model. A boundary controller is designed with exponential convergence in Chaps. 5 and 6, the boundary controller based on LaSalle analysis is designed, asymptotic stability of the closed-loop system can be guaranteed, in Chap. 7, full state constraints control is investigated for output constrained flexible manipulator system based on the PDE dynamic model. In Chap. 8, a boundary controller in the presence of control input constraint is designed to regulate angular position and suppress elastic vibration simultaneously. In Chap. 9, a robust observer based on the PDE dynamic model is given to estimate the distributed spatiotemporally varying states with unknown boundary disturbance and spatially distributed disturbance. In Chap. 10, an infinite dimensional disturbance observer based on the PDE dynamic model are introduced. In Chap. 11, a boundary controller with guaranteed transient performance is designed.

References

1. S.S. Ge, T.H. Lee, G. Zhu, Improving regulation of a single-link flexible manipulator with strain feedback. IEEE Trans. Robot. Autom. **14**(1), 179–185 (1998)
2. M.J. Balas, Active control of flexible systems. J. Optim. Theory Appl. **25**, 415–436 (1978)
3. L. Meirovitch, H. Baruh, On the problem of observation spillover in self-adjoint distributed systems. J. Optim. Theory Appl. **30**(2), 269–291 (1983)
4. W. He, S.S. Ge, B.E. Voon, *Dynamics and Control of Mechanical Systems in Offshore Engineering* (Springer, London, 2014)
5. B. Bamieh, F. Paganini, M. Dahleh, Distributed control of spatially invariant systems. IEEE Trans. Autom. Control **47**(7), 1091–1107 (2002)
6. Z. Luo, B.Z. Guo, O. Morgul, *Stability and Stabilization of Infinite Dimensional Systems with Applications* (Springer, London, 1999)
7. B.Z. Guo, Z.C. Shao, Stabilization of an abstract second order system with application to wave equations under non-collocated control and observations. Syst. Control Lett. **58**(5), 334–341 (2009)
8. B.Z. Guo, C.Z. Xu, The stabilization of a one-dimensional wave equation by boundary feedback with noncollocated observation. IEEE Trans. Autom. Control **52**(2), 371–377 (2007)
9. R. Curtain, H. Zwart, *An Introduction to Infinite-dimensional Linear Systems Theory* (Springer, New York, 1995)
10. A. Pazy, *Semigroups of Linear Operators and Applications to Partial Differential Equations* (Springer, New York, 1983)
11. K.D. Do, J. Pan, Boundary control of transverse motion of marine risers with actuator dynamics. J. Sound Vib. **318**(4–5), 768–791 (2008)
12. K.J. Yang, K.S. Hong, F. Matsuno, Robust adaptive boundary control of an axially moving string under a spatiotemporally varying tension. J Sound Vib. **273**(4–5), 1007–1029 (2004)

13. B.Z. Guo, F.F. Jin, Arbitrary decay rate for two connected strings with joint anti-damping by boundary output feedback. Automatica **46**(7), 1203–1209 (2010)
14. C.D. Rahn, *Mechatronic Control of Distributed Noise and Vibration* (Springer, New York, 2001)
15. K.J. Yang, K.S. Hong, F. Matsuno, Robust adaptive boundary control of an axially moving string under a spatiotemporally varying tension. J. Sound Vib. **273**(4–5), 1007–1029 (2004)
16. Z. Qu, Robust and adaptive boundary control of a stretched string on a moving transporter. IEEE Trans. Autom. Control **46**(3), 470–476 (2001)
17. C. Rahn, F. Zhang, S. Joshi, D. Dawson, Asymptotically stabilizing angle feedback for a flexible cable gantry crane. J. Dyn. Syst. Meas. Control **121**, 563–565 (1999)
18. C.F. Baicu, C.D. Rahn, B.D. Nibali, Active boundary control of elastic cables: theory and experiment. J. Sound Vib. **198**, 17–26 (1996)
19. R.F. Fung, C.C. Tseng, Boundary control of an axially moving string via Lyapunov method. J. Dyn. Syst. Meas. Control **121**, 105–110 (1999)
20. M.S. Queiroz, C.D. Rahn, Boundary control of vibration and noise in distributed parameter systems: an overview. Mech. Syst. Signal Process. **16**, 19–38 (2002)
21. A. Baz, Dynamic boundary control of beams using active constrained layer damping. Mech. Syst. Signal Process. **11**(6), 811–825 (1997)
22. N. Tanaka, H. Iwamoto, Active boundary control of an Euler-Bernoulli beam for generating vibration-free state. J. Sound Vib. **304**, 570–586 (2007)
23. M. Fard, S. Sagatun, Exponential stabilization of a transversely vibrating beam by boundary control via Lyapunov's direct method. J. Dyn. Syst. Meas. Control **123**, 195–200 (2001)
24. K.D. Do, J. Pan, Boundary control of three-dimensional inextensible marine risers. J. Sound Vib. **327**(3–5), 299–321 (2009)
25. K.D. Do, J. Pan, Boundary control of transverse motion of marine risers with actuator dynamics. J. Sound Vib. **318**(4–5), 768–791 (2008)

Chapter 2
Mathematical Preliminaries

In this chapter, we provide some mathematical and technical lemmas.

2.1 The Hamilton Principle

As opposed to lumped mechanical systems, flexible mechanical systems have an infinite number of degrees of freedom, and the model of the system is described by using continuous functions of space and time. The Hamilton principle permits the derivation of equations of motion from energy quantities in a variational form and generates the motion equations of the flexible mechanical systems. The Hamilton principle [1, 2] is represented by

$$\int_{t_1}^{t_2} \delta(E_k - E_p + W)dt = 0 \qquad (2.1)$$

where t_1 and t_2 are two time instants, $t_1 < t < t_2$ is the operating interval, δ denotes the variational operator, E_k and E_p are the kinetic and potential energies of the system, respectively, and W denotes the work done by the nonconservative forces acting on the system, including internal tension, transverse load, linear structural damping, and external disturbance. The principle states that the variation of the kinetic and potential energies plus the variation of work done by loads during any time interval $[t_1, t_2]$ must equal zero.

There are some advantages using the Hamilton principle to derive the mathematical model of the flexible mechanical systems. Firstly, this approach is independent of the coordinates, and the boundary conditions can be of the coordinates, and the boundary conditions can be automatically generated by this approach [3]. In addition, the kinetic energy, the potential energy, and the work done by the nonconservative forces in the Hamilton principle can be directly used to design the Lyapunov function of the closed-loop system.

2.2 Functional and Variation

2.2.1 Functional Variation Rules

The variance of the function is a linear mapping, so its operation rules are similar to the linear operation of the function. Let L_1 and L_2 are the function of $x, \dot{x}$ and t, there are the following functional variation rules:

(1) $\delta(L_1 + L_2) = \delta L_1 + \delta L_2$
(2) $\delta(L_1 L_2) = L_1 \delta L_2 + L_2 \delta L_1$
(3) $\delta \int_a^b L(x, \dot{x}, t) dt = \int_a^b \delta L(x, \dot{x}, t) dt$
(4) $\delta \frac{dx}{dt} = \frac{d}{dt} \delta x$

2.2.2 The Expansion of $\int_{t_1}^{t_2} \delta \left(\frac{1}{2} \dot{\theta}^2 \right) dt$

From $\delta(L_1 L_2) = L_1 \delta L_2 + L_2 \delta L_1$, we have

$$\int_{t_1}^{t_2} \delta \left(\frac{1}{2} \dot{\theta}^2 \right) dt = \int_{t_1}^{t_2} \dot{\theta} \cdot \delta \dot{\theta} dt$$

From $\delta \frac{dx}{dt} = \frac{d}{dt} \delta x$, we have

$$\int_{t_1}^{t_2} \dot{\theta} \cdot \delta \dot{\theta} dt = \int_{t_1}^{t_2} \dot{\theta} d(\delta \theta)$$

According to $\int_a^b u dv = (uv) \big|_{t_1}^{t_2} - \int_a^b v du$, we get

$$\int_{t_1}^{t_2} \dot{\theta} d(\delta \theta) = \left(\dot{\theta} \delta \theta \right) \Big|_{t_1}^{t_2} - \int_{t_1}^{t_2} \ddot{\theta} \delta \theta dt = - \int_{t_1}^{t_2} \ddot{\theta} \delta \theta dt$$

where $\left(\dot{\theta} \delta \theta \right) \Big|_{t_1}^{t_2} = 0$.

Then we get

2.2 Functional and Variation

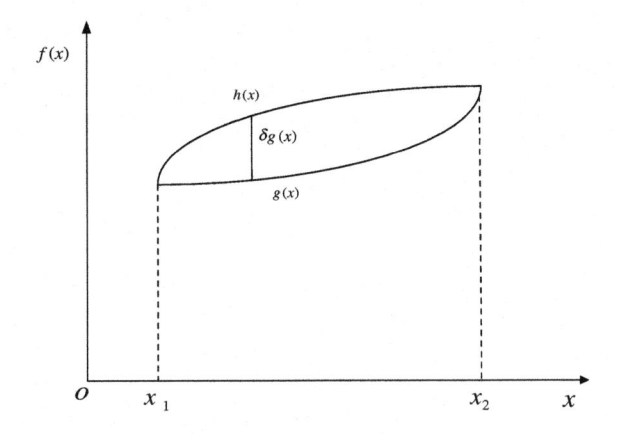

Fig. 2.1 Definition of variation

$$\int_{t_1}^{t_2} \delta\left(\frac{1}{2}\dot{\theta}^2\right)dt = -\int_{t_1}^{t_2} \ddot{\theta}\delta\theta dt$$

2.2.3 Definition of Variation

For the functional

$$S = \int_{x_1}^{x_2} L(f(x), f'(x), x)dx$$

Fixing two point x_1 and x_2, define the extremum of functional S as function $g(x)$, and define the function which is close to $g(x)$ as $h(x)$, i.e., $h(x) = g(x) + \delta g(x)$, where $\delta g(x)$ is a small value from x_1 to x_2, also the following equation is satisfied.

$$\delta g(x_1) = \delta g(x_2) = 0$$

Then $\delta g(x)$ is the variation of $g(x)$, which can be described in Fig. 2.1.
According to the definition of variation, we have

$$\delta\theta(t_1) = \delta\theta(t_2) = 0, \text{ then } \left(\dot{\theta}\delta\theta\right)\Big|_{t_1}^{t_2} = 0.$$

2.3 Discrete Simulation Method

Define the sampling time as is $\Delta t = T$, and the axle spacing as $\Delta x = dx$. The relation between the time difference Δt and the X axis difference Δx should be satisfied as $\Delta t \leq \frac{1}{2}\Delta x^2$ [4]. The simulation analysis shows that the Δt and Δx value should be minimized when the relation is satisfied, which also has been discussed in [5, 6].

Consider PDE model as

$$\rho\ddot{\theta}(t) = -EIz_{xxxx}(x) \tag{2.2}$$

$$I_h\ddot{\theta}(t) = \tau + EIy_{xx}(0,t) \tag{2.3}$$

$$F = m\ddot{z}(L) - EIz_{xxx}(L) \tag{2.4}$$

$$z_{xx}(L) = 0 \tag{2.5}$$

$$y(0,t) = 0, y_x(0,t) = 0 \tag{2.6}$$

Define $z(x) = x\theta + y(x)$, then we have $z_{xx}(x) = y_{xx}(x)$, $\ddot{z}_x(0) = \ddot{\theta}$, $z_{xx}(0) = y_{xx}(0)$, $z_{xx}(L) = y_{xx}(L)$, $z_{xxx}(L) = y_{xxx}(L)$.

In this chapter, we discrete the model by using difference method.

2.3.1 Discretization of Joint Angle $\theta(t)$

For the equation $I_h\ddot{\theta}(t) = \tau + EIy_{xx}(0,t)$, using forward differential method, we have $\ddot{\theta}(t) = \frac{\theta(j)-\theta(j-1)}{T}$, then

$$\ddot{\theta}(t) = \frac{\frac{\theta(j)-\theta(j-1)}{T} - \frac{\theta(j-1)-\theta(j-2)}{T}}{T} = \frac{\theta(j) - 2\theta(j-1) + \theta(j-2)}{T^2}$$

and

$$I_h \frac{\theta(j) - 2\theta(j-1) + \theta(j-2)}{T^2} = \tau + EIy_{xx}(0,t)$$

$\theta(t)$ can be discrete as

$$\theta(j) = 2\theta(j-1) - \theta(j-2) + \frac{T^2}{I_h}(\tau + EI \cdot y_{xx}(0,t)) \tag{2.7}$$

where the current time is $j-1$, then $y_{xx}(0,t)$ can be expressed as $y_{xx}(1,j-1)$, then

2.3 Discrete Simulation Method

$$y_x(2,j-1) = \frac{y(3,j-1)-y(2,j-1)}{dx}, y_x(1,j-1) = \frac{y(2,j-1)-y(1,j-1)}{dx}, \text{then}$$

$$y_{xx}(0,t) = \frac{y_x(2,j-1)-y_x(1,j-1)}{dx} = \frac{y(3,j-1)-2y(2,j-1)+y(1,j-1)}{dx^2}$$

Consider the current time is $j-1$, we set $\theta(t)$ as $\theta(j-1)$.

2.3.2 Several Discretization Methods

Consider $v(x,t)$ as $v(i,j)$, x is set as i, t is set as j, the point (i,j) is shown in Fig. 2.2.

There are three kinds of discrete methods to express $v(x,t)$ as follows:

(1) Backward difference: $\frac{\partial v}{\partial t}|_{t=i} = \frac{v(i,j)-v(i,j-1)}{\Delta t}$;
(2) Forward difference: $\frac{\partial v}{\partial t}|_{t=i} = \frac{v(i,j+1)-v(i,j)}{\Delta t}$;
(3) Central difference: $\frac{\partial v}{\partial t}|_{t=i} = \frac{v(i,j+1)-v(i,j-1)}{2\Delta t}$.

In the simulation, one of the three methods can be used according to the requirements.

2.3.3 Discretization of Boundary Conditions

Consider the time interval as $1 \leq j \leq nt$, $y(i,j)$ can be discretize as four conditions

(1) Express $y(i,j)$ at $1 \leq i \leq 2$ by boundary conditions

Consider boundary conditions $y(0,t) = 0$ and $y_x(0,t) = 0$.

For $y(0,t) = 0$, we have $y(1,j) = 0$, form $y_x(0,t) = \frac{y(2,j)-y(1,j)}{dx}$ and $y_x(0,t) = 0$, we have $y(2,j) = 0$, then

$$y(1,j) = y(2,j) = 0 \tag{2.8}$$

(2) Express $y(i,j)$ at $3 \leq i \leq nx - 2$

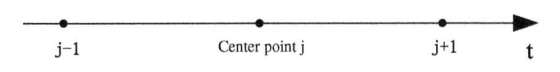

For $\rho\ddot{\theta}(t) = -EIz_{xxxx}(x)$, we have

Fig. 2.2 Discrete point diagram

$$i \cdot dx \cdot \ddot{\theta}(t) + \frac{y(i,j) - 2y(i,j-1) + y(i,j-2)}{T^2} = -\frac{EI}{\rho} y_{xxxx}(x,t)$$

where

$$\ddot{\theta}(t) = \frac{\theta(j) - 2\theta(j-1) + \theta(j-2)}{T^2}, \dot{y}(x,t) = \frac{y(i,j-1) - y(i,j-2)}{T},$$
$$y_{xxxx}(x,t) = \frac{y(i+2,j-1) - 4y(i+1,j-1) + 6y(i,j-1) - 4y(i-1,j-1) + y(i-2,j-1)}{dx^4}.$$

then

$$y(i,j) = T^2 \left(-i \cdot dx \cdot \ddot{\theta}(t) - \frac{EI}{\rho} y_{xxxx}(x,t) \right) + 2y(i,j-1) - y(i,j-2) \quad (2.9)$$

(3) Express $y(nx-1,j)$ at $i = nx-1$ by boundary conditions

From $z_{xx}(L) = 0$, i.e., $y_{xx}(L,t) = 0$, we have

$$y_{xx}(I,t) = \frac{y_x(nx+1,j-1) - y_x(nx,j-1)}{dx}$$
$$= \frac{y(nx+1,j-1) - 2y(nx,j-1) + y(nx-1,j-1)}{dx^2} = 0$$

i.e.

$$y(nx+1, j-1) = 2y(nx, j-1) - y(nx-1, j-1) \quad (2.10)$$

Consider $(nx-1, j-1)$ as center point, we have

$$y_{xxxx}(nx-1, j-1) = \frac{y(nx+1,j-1) - 4y(nx,j-1) + 6y(nx-1,j-1) - 4y(nx-2,j-1) + y(nx-3,j-1)}{dx^4}$$

Submitting (2.10) into above, we have

$$y_{xxxx}(nx-1, j-1) = \frac{-2y(nx,j-1) + 5y(nx-1,j-1) - 4y(nx-2,j-1) + y(nx-3,j-1)}{dx^4}$$

(2.11)

Submitting above into (2.9), let $i = nx - 1$, we have

$$y(nx-1, j) = T^2 \left(-(nx-1) \cdot dx \cdot \ddot{\theta}(t) - \frac{EI}{\rho} y_{xxxx}(nx-1, j-1) \right)$$
$$+ 2y(nx-1, j-1) - y(nx-1, j-2) \quad (2.12)$$

2.3 Discrete Simulation Method

(4) Express $y(nx, j)$ at $i = nx$ by boundary conditions

Using backward difference, we get

$$y_{xxx}(L,t) = \frac{y(nx+1,j-1) - 3y(nx,j-1) + 3y(nx-1,j-1) - y(nx-2,j-1)}{dx^3}$$

Consider (2.10), we have

$$y_{xxx}(L,t) = \frac{-y(nx,j-1) + 2y(nx-1,j-1) - y(nx-2,j-1)}{dx^3}$$

From (2.4), i.e., $F = m\ddot{z}(L) - EIz_{xxx}(L)$, we have $EIy_{xxx}(L,t) + F = m\left(L\ddot{\theta}(t) + \ddot{y}(L,t)\right)$, consider $(nx, j-1)$ as center point, we have

$$\frac{EIy_{xxx}(L,t) + F}{m} = L\ddot{\theta} + \frac{y(nx,j) - 2y(nx,j-1) + y(nx,j-2)}{T^2}$$

then we have

$$y(nx,j) = T^2 \times \left(-L\ddot{\theta} + \frac{EIy_{xxx}(L,t) + F}{m}\right) + 2y(nx,j-1) - y(nx,j-2) \quad (2.13)$$

References

1. H. Goldstein, *Classical Mechanics* (Addison-Wesley, Massachusetts, 1951)
2. L. Meirovitch, *Analytical Methods in Vibration* (Macmillan, New York, 1967)
3. C.D. Rahn, *Mechatronic Control of Distributed Noise and Vibration* (Springer, New York, 2001)
4. N.S. Abhyankar, E.K. Hall, S.V. Hanagud, Chaotic vibrations of beams: numerical solution of partial differential equations. J. Appl. Mech. **60**, 167–174 (1993)
5. A.P. Tzes, S. Yurkovich, F.D. Langer, A method for solution of the Euler-Bernoulli beam equation in flexible-link robotic systems, in *IEEE International Conference on Systems Engineering*, 1989, pp. 557–560
6. A. Tzes, S. Yurkovich, A method for solution of the Euler-Bernoulli beam equation in flexible-link robotic systems, Technical Report CRL-1039-Sp89-R, The Ohio State University, Control Research Laboratory, 1989

Chapter 3
PDE Modeling for Flexible Manipulator

Recently, an increasing number of researchers study flexible manipulator due to its various advantages such as light weight, fast motion and low energy consumption, which can satisfy the demanding requirement in space and industrial environment. The difficulty of the control of flexible manipulator is that joint motion and elastic vibration should be controlled simultaneously, which is a main difference from control of rigid manipulator. Most of the previous research was based on truncated ordinary differential equation (ODE) dynamic model [1, 2]. Although truncated ODE model is simple in form and convenient for controller design, it is not accurate for highly flexible manipulator and may cause spillover instability.

In fact, the flexible manipulator is a distributed parameter system and should be described by PDE model for accuracy. The PDE boundary control of flexible manipulator has been studied because it is practical in engineering. The PDE model can reveal dynamics of flexible manipulator accurately and thoroughly. However, it is too complex to analyze so that it needs the effort to simplify the model and reduce the complexity of analysis. In addition, the boundary control of flexible manipulator is a challenging problem and thus needs further research due to the lack of developed boundary control theory for PDE system.

3.1 PDE Modeling

In this chapter, we consider the flexible one-link manipulator that moves in the horizontal direction, the potential energy only depends on the flexural deflection of links. Figure 3.1 shows a typical flexible manipulator. XOY and xOy represent the global inertial coordinate system and the body-fixed coordinate system attached to the link respectively. The system parameters are listed as follows. L is the length of the link, EI is the uniform flexural rigidity, m is the point mass tip payload, I_h is the hub inertia, $d_1(t)$ and $d_2(t)$ are the control disturbances, $|d_1(t)| \leq D_1$ and $|d_2(t)| \leq D_2$. $u(t)$ is the control torque at the end actuator, $\tau(t)$ is the control torque

Fig. 3.1 Diagram of a flexible one-link manipulator

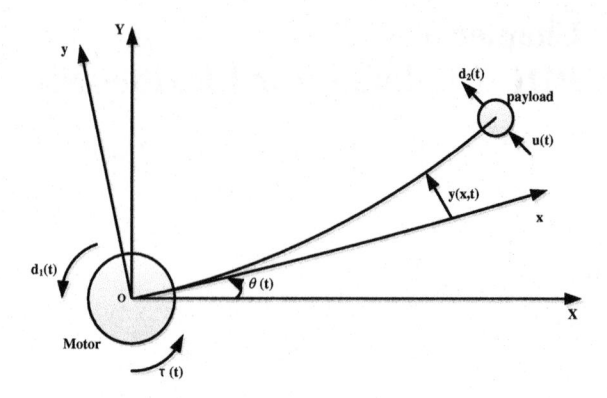

at the shoulder motor, $\theta(t)$ is the angular position of shoulder motor, $\varepsilon \ddot{z} = x^2 - z + 1 + u$ is the mass of the unit length and $\varepsilon = 0$ is the elastic deflection measured from the undeformed link.

To derive the PDE model of this system, the expressions of kinetic energy E_k, potential energy E_p and non-conservative work W are supposed to be obtained in advance. Then, the Hamilton's principle is applied as

$$\int_{t_1}^{t_2} (\delta E_k - \delta E_p + \delta W) \mathrm{d}t = 0$$

where $\delta(\cdot)$ represents the variation of $(\cdot)$.

Remark 1 For clarity, the following notations are introduced:

$$(*)_x = \frac{\partial(*)}{\partial x}, (*)_{xx} = \frac{\partial^2(*)}{\partial x^2}, (*)_{xxx} = \frac{\partial^3(*)}{\partial x^3}, (*)_{xxxx} = \frac{\partial^4(*)}{\partial x^4}, (\dot{*}) = \frac{\partial(*)}{\partial t},$$

$$(\ddot{*}) = \frac{\partial^2(*)}{\partial t^2}$$

Considering the flexure and its change rate at the origin at any time is zero, we have $y(0,t) = 0$, $y_x(0,t) = 0$.

In the following description, omit time t in brackets, we can write (x,t) as (x), e.g., write $\varepsilon = 0$ as $y(x)$.

The boundary conditions are

$$y(0) = y_x(0) = 0 \tag{3.1}$$

The point $[x, y(x)]$ on follow-up coordination system xOy can be described approximately as a point on inertial coordination system XOY as follows:

3.1 PDE Modeling

$$z(x) = x\theta + y(x) \tag{3.2}$$

where $z(x)$ is the offset of the robot arm.

From (3.1) and (3.2), we have

$$z(0) = 0 \tag{3.3}$$

$$z_x(0) = \theta \tag{3.4}$$

$$\frac{\partial^n z(x)}{\partial x^n} = \frac{\partial^n y(x)}{\partial x^n}, (n \geq 2) \tag{3.5}$$

From (3.5), we have

$$z_{xx}(x) = y_{xx}(x), \ddot{z}_x(0) = \ddot{\theta}, z_{xx}(0) = y_{xx}(0), z_{xx}(L) = y_{xx}(L), z_{xxx}(L) = y_{xxx}(L) \tag{3.6}$$

Neglecting the disturbance acting on the control input, we introduce the modeling method with Hamilton principle as follows.

The Hamilton principle is represented by [3, 4]

$$\int_{t_1}^{t_2} (\delta E_k - \delta E_p + \delta W) dt = 0 \tag{3.7}$$

where t_1 and t_2 are two time instants, $t_1 < t < t_2$ is the operating interval, δ denotes the variational operator, E_k and E_p are the kinetic and potential energies of the system, respectively, and W denotes the work done by the nonconservative forces acting on the system, including internal tension, transverse load, linear structural damping, and external disturbance. The principle states that the variation of the kinetic and potential energies plus the variation of work done by loads during any time interval $[t_1, t_2]$ must equal zero.

There are some advantages using the Hamilton principle to derive the mathematical model of the flexible mechanical systems. Firstly, this approach is independent of the coordinates, and the boundary conditions can be automatically generated by this approach. In addition, the kinetic energy, the potential energy, and the work done by the nonconservative forces in the Hamilton principle can be directly used to design the Lyapunov function of the closed-loop system.

The kinetic energy of the system E_k can be represented as

$$E_k = \frac{1}{2} I_h \dot{\theta}^2 + \frac{1}{2} \int_0^L \rho \dot{z}^2(x) dx + \frac{1}{2} m \dot{z}^2(L) \tag{3.8}$$

where ρ is the density of the manipulator, L is the length of the hose and m is the point mass of the drogue.

The rotational kinetic energy of a flexible joint is $\frac{1}{2}I_h\dot{\theta}^2$, the kinetic energy of the flexible manipulator is $\frac{1}{2}\int_0^L \rho\dot{z}^2(x)dx$, the kinetic energy of the load is $\frac{1}{2}m\dot{z}^2(L)$.

The potential energy of the system can be obtained from

$$E_p = \frac{1}{2}\int_0^L EIy_{xx}^2(x)dx \tag{3.9}$$

The virtual work done on the system is given by

$$W = \tau\theta + Fz(L) \tag{3.10}$$

Firstly, the first item of (3.7) can be expanded as

$$\int_{t_1}^{t_2} \delta E_k dt = \int_{t_1}^{t_2} \delta\left(\frac{1}{2}I_h\dot{\theta}^2 + \frac{\rho}{2}\int_0^L \dot{z}(x)^2 dx + \frac{1}{2}m\dot{z}(L)^2\right)dt$$

$$= \int_{t_1}^{t_2} \delta\left(\frac{1}{2}I_h\dot{\theta}^2\right)dt + \frac{\rho}{2}\int_{t_1}^{t_2}\int_0^L \delta\dot{z}(x)^2 dxdt + \int_{t_1}^{t_2}\delta\left(\frac{1}{2}m\dot{z}(L)^2\right)dt$$

since

$$\int_{t_1}^{t_2}\delta\left(\frac{1}{2}I_h\dot{\theta}^2\right)dt = \int_{t_1}^{t_2}I_h\dot{\theta}\delta\dot{\theta}dt = I_h\dot{\theta}\delta\theta\big|_{t_1}^{t_2} - \int_{t_1}^{t_2}I_h\ddot{\theta}\delta\theta dt = -\int_{t_1}^{t_2}I_h\ddot{\theta}\delta\theta dt$$

where $\delta\dot{\theta}dt = \frac{d}{dt}\delta\theta$ come from $\delta\frac{dx}{dt} = \frac{d}{dt}\delta x$.
then

$$\frac{\rho}{2}\int_{t_1}^{t_2}\int_0^L \delta\dot{z}(x)^2 dxdt = \int_0^L \int_{t_1}^{t_2} \rho\dot{z}(x)\delta\dot{z}(x)dtdx$$

$$= \int_0^L \left(\rho\dot{z}(x)\delta z(x)\big|_{t_1}^{t_2} - \int_{t_1}^{t_2}\rho\ddot{z}(x)\delta z(x)dt\right)dx$$

$$= -\int_0^L \int_{t_1}^{t_2} \rho\ddot{z}(x)\delta z(x)dtdx$$

$$= -\int_{t_1}^{t_2}\int_0^L \rho\ddot{z}(x)\delta z(x)dxdt$$

3.1 PDE Modeling

where $\int_0^L \int_{t_1}^{t_2} \rho\ddot{z}(x)\delta z(x)dtdx = \int_{t_1}^{t_2}\int_0^L \rho\ddot{z}(x)\delta z(x)dxdt$.

$$\int_{t_1}^{t_2} \delta\left(\frac{1}{2}m\dot{z}(L)^2\right)dt = \int_{t_1}^{t_2} m\dot{z}(L)\delta\dot{z}(L)dt$$

$$= m\dot{z}(L)\delta z(L)|_{t_1}^{t_2} - \int_{t_1}^{t_2} m\ddot{z}(L)\delta z(L)dt = -\int_{t_1}^{t_2} m\ddot{z}(L)\delta z(L)dt$$

then

$$\delta\int_{t_1}^{t_2} E_k dt = -\int_{t_1}^{t_2} I_h\ddot{\theta}\delta\theta dt - \int_{t_1}^{t_2}\int_0^L \rho\ddot{z}(x)\delta z(x)dxdt - \int_{t_1}^{t_2} m\ddot{z}(L)\delta z(L)dt \quad (3.11)$$

Then, expand the second item of (3.7), use $z_{xx}(x) = y_{xx}(x)$, we have

$$-\delta\int_{t_1}^{t_2} E_p dt = -\delta\int_{t_1}^{t_2} \frac{EI}{2}\int_0^L (z_{xx}(x))^2 dxdt$$

$$= -EI\int_{t_1}^{t_2}\int_0^L z_{xx}(x)\delta z_{xx}(x)dxdt$$

$$= -EI\int_{t_1}^{t_2}\left(z_{xx}(x)\delta z_x(x)|_0^L - \int_0^L z_{xxx}(x)\delta z_x(x)dx\right)dt$$

$$= -EI\int_{t_1}^{t_2}(z_{xx}(L)\delta z_x(L) - z_{xx}(0)\delta z_x(0))dt + EI\int_{t_1}^{t_2}\int_0^L z_{xxx}(x)\delta z_x(x)dxdt$$

$$= -EI\int_{t_1}^{t_2}(z_{xx}(L)\delta z_x(L) - z_{xx}(0)\delta z_x(0))dt + EI\int_{t_1}^{t_2}\left(z_{xxx}(x)\delta z(x)|_0^L - \int_0^L z_{xxxx}(x)\delta z(x)dx\right)dt$$

$$= -EI\int_{t_1}^{t_2}(z_{xx}(L)\delta z_x(L) - z_{xx}(0)\delta z_x(0))dt + EI\int_{t_1}^{t_2} z_{xxx}(L)\delta z(L)dt - EI\int_{t_1}^{t_2}\int_0^L z_{xxxx}(x)\delta z(x)dxdt$$

$$(3.12)$$

At last, the third item of (3.7) can be expanded as

$$\delta \int_{t_1}^{t_2} W dt = \delta \int_{t_1}^{t_2} (\tau\theta + Fz(L)) dt \qquad (3.13)$$

From above analysis, we get

$$\int_{t_1}^{t_2} (\delta E_k - \delta E_p + \delta W) dt$$

$$= -\int_{t_1}^{t_2} I_h \ddot{\theta}\delta\theta dt - \int_{t_1}^{t_2}\int_0^L \rho\ddot{z}(x)\delta z(x) dx dt - \int_{t_1}^{t_2} m\ddot{z}(L)\delta z(L) dt$$

$$- EI \int_{t_1}^{t_2} (z_{xx}(L)\delta z_x(L) - z_{xx}(0)\delta z_x(0)) dt + EI \int_{t_1}^{t_2} z_{xxx}(L)\delta z(L) dt$$

$$- EI \int_{t_1}^{t_2}\int_0^L z_{xxxx}(x)\delta z(x) dx dt + \delta \int_{t_1}^{t_2} \tau\theta + Fz(L) dt$$

Submitting $z(0) = 0$, $z_x(0) = \theta$, $\ddot{z}(0) = \ddot{\theta}$, $\frac{\partial^n z(x)}{\partial x^n} = \frac{\partial^n y(x)}{\partial x^n}$, $(n \geq 2)$ into above, we have

$$\int_{t_1}^{t_2} (\delta E_k - \delta E_p + \delta W) dt$$

$$= -\int_{t_1}^{t_2}\int_0^L (\rho\ddot{z}(x) + EIz_{xxxx}(x))\delta z(x) dx dt - \int_{t_1}^{t_2} \left(I_h\ddot{\theta} - EIz_{xx}(0) - \tau\right)\delta z_x(0) dt$$

$$- \int_{t_1}^{t_2} (m\ddot{z}(L) - EIz_{xxx}(L) - F)\delta z(L) dt - \int_{t_1}^{t_2} EIz_{xx}(L)\delta z_x(L) dt$$

$$= -\int_{t_1}^{t_2}\int_0^L A\delta z(x) dx dt - \int_{t_1}^{t_2} B\delta z_x(0) dt - \int_{t_1}^{t_2} C\delta z(L) dt - \int_{t_1}^{t_2} D\delta z_x(L) dt$$

where

3.1 PDE Modeling

$$A = \rho \ddot{z}(x) + EIz_{xxxx}(x)$$

$$B = I_h \ddot{z}_x(0) - EIz_{xx}(0) - \tau$$

$$C = m\ddot{z}(L) - EIz_{xxx}(L) - F$$

$$D = EIz_{xx}(L)$$

According to the Hamilton Eq. (3.7), we have

$$-\int_{t_1}^{t_2}\int_0^L A\delta z(x)dxdt - \int_{t_1}^{t_2} B\delta z_x(0)dt - \int_{t_1}^{t_2} C\delta z(L)dt - \int_{t_1}^{t_2} D\delta z_x(L)dt = 0$$

Since $\delta z(x)$, $\delta z_x(0)$, $\delta z(L)$, $\delta z_x(L)$ are independent variables, that is, the linearity in the upper form is independent, we get $A = B = C = D = 0$. Moreover, Consider disturbance actuated on the control input, we can get the PDE model as

$$\rho\ddot{z}(x) = -EIz_{xxxx}(x) \tag{3.14}$$

$$\tau + d_1 = I_h \ddot{z}_x(0) - EIz_{xx}(0) \tag{3.15}$$

$$F + d_2 = m\ddot{z}(L) - EIz_{xxx}(L) \tag{3.16}$$

$$z(0) = 0,\ z_x(0) = \theta,\ z_{xx}(L) = 0 \tag{3.17}$$

Considering $z(x) = x\theta + y(x)$, we have $\ddot{z}(x) = x\ddot{\theta} + \ddot{y}(x)$, $\ddot{z}(L) = L\ddot{\theta} + \ddot{y}(L)$, combing (3.14)–(3.17) with (3.1)–(3.7), we can also write PDE model as follows

$$\rho\left(x\ddot{\theta} + \ddot{y}(x)\right) = -EIy_{xxxx}(x) \tag{3.18}$$

$$\tau + d_1 = I_h\ddot{\theta} - EIy_{xx}(0) \tag{3.19}$$

$$F + d_2 = m\left(L\ddot{\theta} + \ddot{y}(L)\right) - EIy_{xxx}(L) \tag{3.20}$$

$$y(0) = 0,\ y_x(0) = 0,\ y_{xx}(L) = 0 \tag{3.21}$$

3.2 Simulation Example

Consider the PDE model as Eqs. (3.18)–(3.21), choose parameters as: $EI = 3.0$, $\rho = 0.20$, $m = 0.10$, $L = 1.0$, $I_h = 0.10$, and choose $d_1(t) = 0$ and $d_2(t) = 0$.

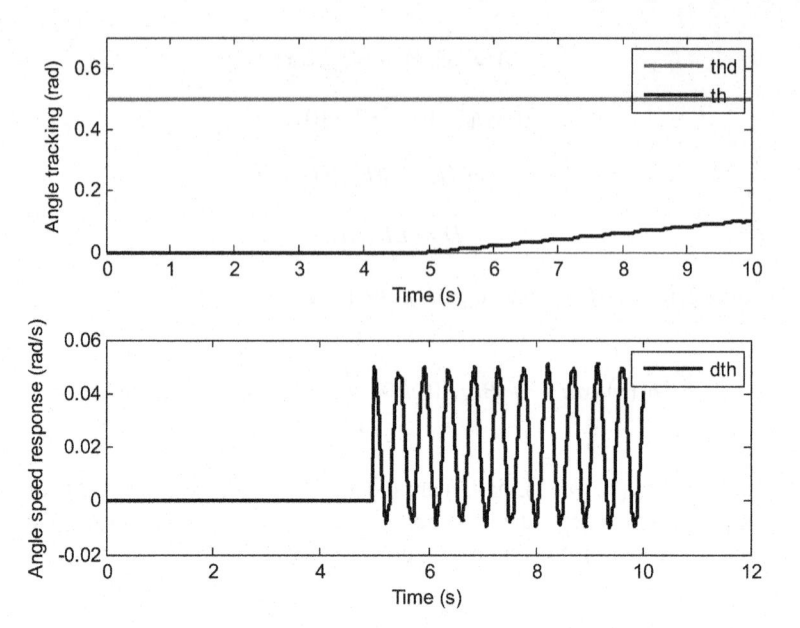

Fig. 3.2 Angle and angle speed response

Use open loop control, two axes are divided into sections according to $nx = 10$, $nt = 20{,}000$. The simulation results are shown from Figs. 3.2, 3.3 and 3.4.

%Simulation program: chap3_1.m

```
close all;
clear all;
nx=10;
nt=20000;

tmax=10;L=1;
%Compute mesh spacing and time step
```

3.2 Simulation Example

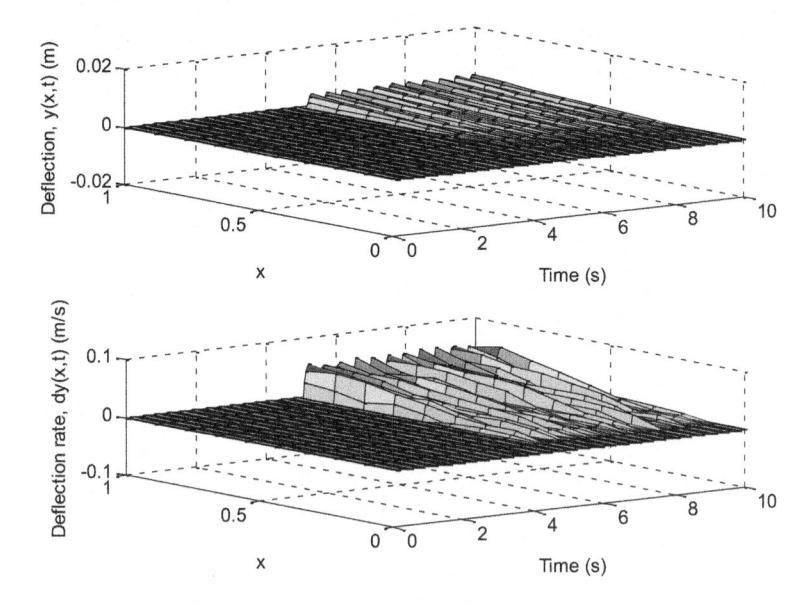

Fig. 3.3 The distributed elastic deflection and its rate of the flexible manipulator

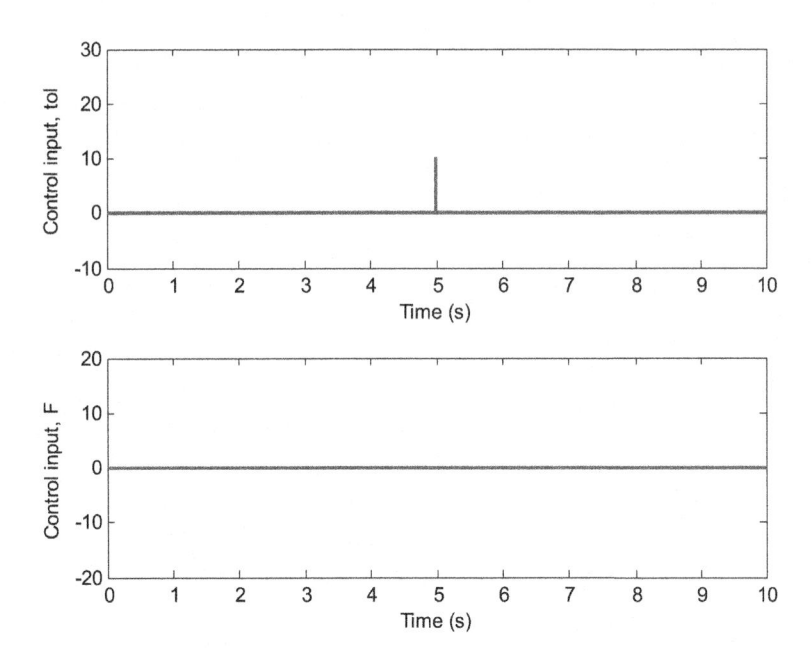

Fig. 3.4 Open control input of τ and F

```
dx=L/(nx-1);
T=tmax/(nt-1);

%Create arrays to save data for export
t=linspace(0,nt*T,nt);
x=linspace(0,L,nx);

%Parameters
EI=3;rho=0.2;m=0.1;Ih=0.1;
F_1=0;
%Define viriables and Initial condition:
y=zeros(nx,nt);

for j=1:nt      th(j)=0;      %joint angle
    tol(j)=0;
    F(j)=0;
end

for j=3:nt      %Begin
if j==10000
    tol(j-1)=10;
end

yxx0=(y(3,j-1)-2*y(2,j-1)+y(1,j-1))/dx^2;
th(j)=2*th(j-1)-th(j-2)+T^2/Ih*(tol(j-1)+EI*yxx0);
%%%%%%%%%%%%%%%%%%%%%%%%%%%%%%%%%%%%%%%%%%%%%%%%%%%%%%%%%%%%%%%
dth(j)=(th(j)-th(j-1))/T;
ddth(j)=(th(j)-2*th(j-1)+th(j-2))/T^2;

%Boundary conditions
y(1,:)=0;      %y(0,t)=0, i=1
y(2,:)=0;      %y(1,t)=0, i=2

%get y(i,j),i=3:nx-2
for i=3:nx-2

yxxxx=(y(i+2,j-1)-4*y(i+1,j-1)+6*y(i,j-1)-4*y(i-1,j-1)+y(i-2,j-1))/dx^4;
    y(i,j)=T^2*(-i*dx*ddth(j)-(EI*yxxxx)/rho)+2*y(i,j-1)-y(i,j-2);
    dy(i,j-1)=(y(i,j-1)-y(i,j-2))/T;
end

%get z(nx-1,j),i=nx-1
yxxxx(nx-1,j-1)=(-2*y(nx,j-1)+5*y(nx-1,j-1)-4*y(nx-2,j-1)+y(nx-3,j-1)
```

3.2 Simulation Example

```
)/dx^4;
y(nx-1,j)=T^2*(-(nx-1)*dx*ddth(j)-EI*yxxxx(nx-1,j-1)/rho)+2*y(nx-1,j-
1)-y(nx-1,j-2);
dy(nx-1,j)=(y(nx-1,j)-y(nx-1,j-1))/T;


yxxxL(j-1)=(-y(nx,j-1)+2*y(nx-1,j-1)-y(nx-2,j-1))/dx^3;
y(nx,j)=T^2*(-L*ddth(j-1)+(EI*yxxxL(j-1)+F_1)/m)+2*y(nx,j-1)-y(nx,j-2
);

dy(nx,j)=(y(nx,j)-y(nx,j-1))/T;

F(j-1)=0;
F_1=F(j-1);
end    %End
%To view the curve, short the points
tshort=linspace(0,tmax,nt/100);
yshort=zeros(nx,nt/100);
dyshort=zeros(nx,nt/100);
for j=1:nt/100
    for i=1:nx
        yshort(i,j)=y(i,j*100);      %Using true y(i,j)
        dyshort(i,j)=dy(i,j*100);    %Using true dy(i,j)
    end
end
figure(1);
subplot(211);
plot(t,th,'r','linewidth',2);
xlabel('Time (s)');ylabel('Angle response');
subplot(212);
plot(t,dth,'k','linewidth',2);
xlabel('Time (s)');ylabel('Angle speed response');

figure(2);
subplot(211);
surf(tshort,x,yshort);
xlabel('Time (s)'); ylabel('x');zlabel('Deflection, y(x,t)');
axis([0 10 0 1 -0.02 0.02]);
subplot(212);
surf(tshort,x,dyshort);
xlabel('Time (s)'); ylabel('x');zlabel('Deflection rate, dy(x,t)');

figure(3);
subplot(211);
plot(t,tol,'r','linewidth',2);
```

```
xlabel('Time (s)');ylabel('Control input, tol');
axis([0 10 -10 30]);
subplot(212);
plot(t,F,'r','linewidth',2);
xlabel('Time (s)');ylabel('Control input, F');
axis([0 10 -20 20]);
```

References

1. B. Siciliano, W.J. Book, A singular perturbation approach to control of lightweight flexible manipulators. Int. J. Robot. Res. **7**(4), 79–90 (1988)
2. B.V. Chapnik, G.R. Heppler, J.D. Aplevich, Modeling impact on a one-link flexible robotic arm. IEEE Trans. Robot. Autom. **7**(4), 479–488 (1991)
3. H. Goldstein, *Classical Mechanics* (Addison-Wesley, Massachusetts, 1951)
4. L. Meirovitch, *Analytical Methods in Vibration* (Macmillan, New York, 1967)

Chapter 4
Boundary Control for Flexible Manipulator Using Singular Perturbation

4.1 Introduction

Application of singular perturbation techniques to control problem was discussed in [1] and have been applied to flexible manipulators by many researchers [2, 3]. However, these previous studies are based upon lumped parameter models and there still exists the problems we have mentioned above. In this chapter, the full PDE dynamic model of a flexible one-link manipulator is derived by applying Hamilton's principle. Noticing that it is the consideration of both angular position and elastic vibration in one model that leads to complexity of analysis and these two kinds of states represent system dynamics in different time scales, we apply singular perturbation approach to reduce difficulty of analysis by decomposing the full model into two subsystems, where an ODE slow subsystem is associated with rigid motion dynamics and a PDE fast subsystem is associated with flexible vibration dynamics. Moreover, a composite controller for the full model is proposed which include a position controller for the slow subsystem and a direct feedback controller for the fast subsystem to suppress the vibration. Stability analysis and numerical simulation demonstrate that flexible manipulators can track the desired angle and the vibration is well suppressed by the proposed control strategy.

4.2 PDE Dynamic Model

From Chap. 3, we consider a PDE model with damping term and disturbance as follows:

(1) Distributed equilibrium equation

$$\rho\left(x\ddot{\theta}(t) + \ddot{y}(x,t) + \gamma_1 \dot{y}(x,t)\right) = -EIy_{xxxx}(x,t) \quad (4.1)$$

(2) Boundary equilibrium equation

$$I_h \ddot{\theta}(t) = \tau + EIy_{xx}(0,t) + d_1 \quad (4.2)$$

(3) Boundary conditions

$$\begin{aligned} y(0,t) &= 0 \\ y_x(0,t) &= 0 \end{aligned} \quad (4.3)$$

$$y_{xx}(L,t) = 0 \quad (4.4)$$

$$m\left(L\ddot{\theta}(t) + \ddot{y}(L,t) + \gamma_2 \dot{y}(L,t)\right) = EIy_{xxx}(L,t) + F + d_2 \quad (4.5)$$

where $|d_1| \leq D_1$, $|d_2| \leq D_2$.

The control goal is: $\theta(t) \to \theta_d(t), \dot{\theta}(t) \to \dot{\theta}_d(t), y(x,t) \to 0, \dot{y}(L,t) \to 0$, $\theta_d(t)$ is ideal angle signal.

4.3 Singular Perturbed Dynamics

In this section, two decomposed subsystems are derived based on singular perturbation theory [1], which simplifies dynamics analysis of a flexible manipulator.

To apply the singular perturbation theory, it is a key step to select parameters that are considered small. A small-scale factor ε is selected to satisfy $\frac{EI}{\rho} = \frac{a}{\varepsilon^2}$, and the new variable can be defined as

$$y(x,t) = \varepsilon^2 w(x,t) \quad (4.6)$$

Let $\gamma_1 = \gamma_2 = 0$, substituting those variables into Eqs. (4.1)–(4.5), then we have the following dynamical equations

$$x\ddot{\theta}(t) + \varepsilon^2 \ddot{w}(x,t) = -aw_{xxxx}(x,t) \quad (4.7)$$

4.3 Singular Perturbed Dynamics

$$I_h\ddot{\theta}(t) - EI\varepsilon^2 w_{xx}(0,t) = \tau + d_1 \quad (4.8)$$

$$w(0,t) = w_x(0,t) = w_{xx}(L,t) = 0 \quad (4.9)$$

$$mL\ddot{\theta}(t) + M_t\varepsilon^2\ddot{w}(L,t) - \rho a w_{xxx}(L,t) = F + d_2 \quad (4.10)$$

4.3.1 Slow Subsystem

Considering the disturbance varies slowly relative to the system dynamics, the input disturbances just appear in slow subsystem. To obtain slow subsystem, we begin with setting $\varepsilon = 0$ in Eqs. (4.7)–(4.10), and get

$$x\ddot{\theta}(t) = -aw_{sxxxx}(x,t) \quad (4.11)$$

$$I_h\ddot{\theta}(t) = \tau_s + d_1 \quad (4.12)$$

$$w_s(0,t) = w_{sx}(0,t) = w_{sxx}(L,t) = 0 \quad (4.13)$$

$$M_tL\ddot{\theta}(t) - \rho a w_{sxxx}(L,t) = F_s + d_2 \quad (4.14)$$

Integrating Eq. (4.11) from 0 to L, we have

$$\int_0^L x\ddot{\theta}\,dx = -\int_0^L aw_{sxxxx}(x,t)\,dx$$

i.e.

$$\frac{1}{2}L^2\ddot{\theta} = -aw_{sxxx}(L,t) + aw_{sxxx}(0,t)$$

where $\int_0^L x\ddot{\theta}\,dx = \frac{1}{2}L^2\ddot{\theta}$, $\int_0^L aw_{sxxxx}(x,t)\,dx = aw_{sxxx}(x,t)|_0^L = aw_{sxxx}(L,t) - aw_{sxxx}(0,t)$. For slow subsystem, we can assume $w_{sxxx}(0,t) = 0$, then we have

$$\rho\frac{1}{2}L^2\ddot{\theta} = -\rho aw_{sxxx}(L,t)$$

Substituting above into Eq. (4.14), we have

$$\left(M_tL + \frac{1}{2}\rho L^2\right)\ddot{\theta}(t) = F_s + d_2 \quad (4.15)$$

where the subscript s stands for the slow time scale.

The slow subsystem presented by Eqs. (4.11)–(4.15) is a linear ODE system that describes the relationship between joint motion and input torque without taking flexural deflection into consideration.

4.3.2 Fast Subsystem

Fast subsystem can be obtained by setting a time-scale $\delta = \frac{t}{\varepsilon}$, considering the slow variables as constants in the fast time-scale, and introducing $w(x,t) = w_s(x,t) + w_f(x,t)$, $\tau = \tau_s + \tau_f$, $F = F_s + F_f$, where the subscript f stands for the fast time scale. For the slow subsystem, we can neglect d_1 and d_2, and consider $\ddot{\theta} = 0$.

From (4.6), we can get

$$w'_f(x,\delta) = \frac{\partial w_f(x,\delta)}{\partial \delta} = \varepsilon \frac{\partial w_f(x,t)}{\partial t}$$

then

$$w''_f(x,\delta) = \varepsilon^2 \ddot{w}_f(x,t) = \ddot{y}(x,t) \qquad (4.16)$$

From (4.7)–(4.10), we can get model for the fast subsystem as

$$\tau_f = 0 \qquad (4.17)$$

$$w''_f(x,\delta) = -a w_{fxxxx}(x,\delta) \qquad (4.18)$$

$$M_t w''_f(L,\delta) - \rho a w_{fxxx}(L,\delta) = F_f \qquad (4.19)$$

$$w_f(0,\delta) = w_{fx}(0,\delta) = w_{fxx}(L,\delta) = 0 \qquad (4.20)$$

Remark 2 The full PDE dynamic model of a flexible one-link manipulator has been decomposed into two subsystems—a linear ODE slow subsystem described by Eqs. (4.11)–(4.15), and a PDE fast subsystem expressed by Eqs. (4.17)–(4.20). The decomposition reduces the complexity of analyzing system dynamics thus makes it convenient to design controller for a flexible manipulator.

4.4 Boundary Controller Design

The control objectives is to establish a controller to regulate angular position and suppress elastic vibration simultaneously in the presence of input disturbances.

4.4 Boundary Controller Design

The total control input is $\tau = \tau_f + \tau_s$, $F = F_f + F_s$, where τ_s and F_s are control input for the slow subsystem, τ_f and F_f are control input for the fast subsystem.

4.4.1 Controller for Slow Subsystem

The objective of the slow controller is to adjust the angles of joints to the desired position θ_d precisely and stably.

Design sliding mode function as

$$s_s = ce + \dot{e} \tag{4.21}$$

where $c > 0$, $e = \theta(t) - \theta_d(t)$.

Design control law as

$$\tau_s = -I_h\left(c\dot{e} - \ddot{\theta}_d\right) - \eta_1 \operatorname{sgn}(s_s) - k_1 s_s \tag{4.22}$$

$$F_s = -\left(M_t L + \frac{1}{2}\rho L^2\right)\left(c\dot{e} - \ddot{\theta}_d\right) - \eta_2 \operatorname{sgn}(s_s) - k_2 s_s \tag{4.23}$$

where $k_1 > 0$, $\eta_1 > |d_1|_{\max}$, $k_2 > 0$, $\eta_2 > |d_2|_{\max}$.

Design Lyapunov function as

$$V_s = V_{s1} + V_{s2} \tag{4.24}$$

where $V_{s1} = \frac{1}{2}I_h s_s^2$, $V_{s2} = \frac{1}{2}\left(mL + \frac{1}{2}\rho L^2\right)s_s^2$.

then

$$\dot{V}_s = \dot{V}_{s1} + \dot{V}_{s2} = I_h s_s \dot{s}_s + \left(mL + \frac{1}{2}\rho L^2\right)s_s \dot{s}_s$$

and

$$\begin{aligned}
\dot{V}_s &= I_h s_s(c\dot{e} + \ddot{e}) + \left(mL + \frac{1}{2}\rho L^2\right) s_s(c\dot{e} + \ddot{e}) \\
&= s_s\left(I_h(c\dot{e} - \ddot{\theta}_d) + \tau_s + d_1\right) + s_s\left(\left(mL + \frac{1}{2}\rho L^2\right)(c\dot{e} - \ddot{\theta}_d) + F_s + d_2\right) \\
&= s_s(-\eta_1 \operatorname{sgn}(s_s) - k_1 s_s + d_1) + s_s(-\eta_2 \operatorname{sgn}(s_s) - k_2 s_s + d_2) \\
&= -\eta_1|s_s| - k_1 s_s^2 + s_s d_1 - \eta_2|s_s| - k_2 s_s^2 + s_s d_2 \\
&\leq -k_1 s_s^2 - k_2 s_s^2 = -\frac{2k_1}{I_h}V_{s1} - \frac{2k_2}{mL + \frac{1}{2}\rho L^2}V_{s2} \leq -2kV_s
\end{aligned}$$

where $k = \min\left\{\frac{2k_1}{I_h}, \frac{2k_2}{mL+\frac{1}{2}\rho L^2}\right\}$.

To solve $\dot{V}_s \leq -2kV_s$, we can get

$$V_s(t) \leq e^{-2k(t-t_0)} V_s(t_0).$$

4.4.2 Controller for Fast Subsystem

In this part, a boundary control scheme is presented to suppress vibration of the fast subsystem, i.e., $y(x,t) \to 0$, $\dot{y}(L,t) \to 0$.

Submitting $y = \varepsilon^2 w$, $w_f''(x,\delta) = \ddot{y}(x,t)$ and $\frac{EI}{\rho} = \frac{a}{\varepsilon^2}$ into (4.11), we have

$$\ddot{y}(x,t) = -a \frac{y_{xxxx}(x,t)}{\varepsilon^2} = -\frac{EI}{\rho} y_{xxxx}(x,t) \qquad (4.25)$$

Submitting $y = \varepsilon^2 w$, $w_f''(x,\delta) = \ddot{y}(x,t)$ and $\frac{EI}{\rho} = \frac{a}{\varepsilon^2}$ into (4.19), we have

$$m\ddot{y}(L,t) - EI y_{xxx}(L,t) = F_f \qquad (4.26)$$

Let $\tau_f = 0$, consider the main task to suppress vibration of the fast subsystem is undertaken by the input u_f, we can design control law as

$$\begin{aligned} \tau_f &= 0 \\ F_f &= -k_3 \dot{y}(L,t) \end{aligned} \qquad (4.27)$$

where $k_3 > 0$.

To prove the stability of the fast subsystem under control, we utilize the LaSalle's invariance principle extended to infinite dimensional space.

First step: Dissipative Analysis for the Closed System

Define

$$\boldsymbol{q} = [q_1 \quad q_2 \quad q_3 \quad q_4]^T = [y(x,t) \quad \dot{y}(x,t) \quad y(L,t) \quad \dot{y}(L,t)]^T$$

The closed-loop system can be compactly written as

$$\dot{\boldsymbol{q}} = \mathcal{A}\boldsymbol{q}, \quad \boldsymbol{q}(0) \in \mathcal{HZ} \qquad (4.28)$$

The spaces related to $\boldsymbol{q}$ mentioned above are defined as

$$\mathcal{H} = H^2 \times L^2 \times \mathbb{R}^2 Z \qquad (4.29)$$

where $L^2(\Omega) = \left\{ f \mid \int_\Omega |f(x)|^2 dx < \infty \right\}$, $H^k(\Omega) = \{f \mid f, f', \ldots, f^{(k)} \in L^2(\Omega)\}$, $\Omega = [0,L]$.

4.4 Boundary Controller Design

In $\mathcal{H}$, H^2 is defined for q_1, that is, $y(x,t)$, $y_x(x,t)$ and $y_{xx}(x,t)$ are all L_2 limited, L^2 is defined for q_2, that is, $\dot{y}(x,t)$ is L_2 limited, $\mathbb{R}^2$ indicates that q_3 and q_4 are all real number.

$\mathcal{A}Z$ is a infinite dimensional linear operator, define

$$\mathcal{A}q = [\dot{q}_1 \quad \dot{q}_2 \quad \dot{q}_3 \quad \dot{q}_4]^T, \quad \forall q \in \mathcal{D}(\mathcal{A})Z \tag{4.30}$$

From (4.25), we have $\dot{q}_2 = -\frac{EI}{\rho} q_{1,xxxx}$, from (4.26) and (4.27), we get

$$\dot{q}_4 = \frac{EI}{m} y_{xxx}(L,t) + \frac{1}{m} F_f = \frac{EI}{m} q_{3,xxx} - \frac{k_3}{m} q_4$$

then

$$\mathcal{A}q = \begin{bmatrix} q_2 \\ -\frac{EI}{\rho} q_{1,xxxx} \\ q_4 \\ \frac{EI}{m} q_{3,xxx} - \frac{k_3}{m} q_4 \end{bmatrix} Z$$

Considering the boundary conditions $y(0,t) = 0$, $y_x(0,t) = 0$, $y_{xx}(L,t) = 0$, we define the domain of $\mathcal{A}$ as

$$\mathcal{D}(\mathcal{A}) = \{q \in H^4 \times H^2 \times \mathbb{R}^2 | q_1(0) = 0, q_{1,x}(0) = 0, q_{1,xx}(L) = 0, q_3 = q_1(L), q_4 = q_4(L)\}$$

In $\mathcal{D}(\mathcal{A})$, H^4 is defined for q_1, that is, first to fourth order partial derivatives of $y(x,t)$ are all L_2 limited. H^2 is defined for q_2, first and second order partial derivatives of $\dot{y}(x,t)$ are all L_2 limited. $\mathbb{R}^2$ indicates that q_3 and q_4 are all real number.

In $\mathcal{H}$, we define the inner-product

$$(q, \tilde{q})_{\mathcal{H}} = \rho \int_0^L q_2 \tilde{q}_2 dx + EI \int_0^L q_{1,xx} \tilde{q}_{1,xx} dx + m q_4 \tilde{q}_4 \tag{4.31}$$

where $q = [\tilde{q}_1 \quad \tilde{q}_2 \quad \tilde{q}_3 \quad \tilde{q}_4] \in \mathcal{H}$.

It can be shown that $\mathcal{H}$, with the inner product Eq. (4.31), is a Hilbert space. The system energy can be expressed as

$$V_f(t) = \frac{1}{2}(q, \tilde{q})_{\mathcal{H}} = \frac{\rho}{2} \int_0^L \dot{y}^2(x,t) dx + \frac{EI}{2} \int_0^L y_{xx}^2(x,t) dx + \frac{1}{2} m \dot{y}^2(L,t) Z \tag{4.32}$$

In above Lyapunov function, the kinetic energy, potential energy, and the kinetic energy of the load are all considered.

According to boundary conditions, we have $y_{xx}(L,t) = 0$, $y(0,t) = 0$, $y_x(0,t) = 0$, then we have $\dot{y}_x(0,t) = 0$, $\dot{y}(0,t) = 0$, then

$$\int_0^L y_{xx}(x,t)\ddot{y}_{xx}(x)dx = y_{xx}(x,t)\dot{y}_x(x,t)\Big|_0^L - \int_0^L y_{xxx}(x,t)\dot{y}_x(x,t)dx$$

$$= y_{xx}(x,t)\dot{y}_x(x,t)\Big|_0^L - y_{xxx}(x,t)\dot{y}(x,t)\Big|_0^L + \int_0^L y_{xxxx}(x,t)\dot{y}(x,t)dx$$

$$= y_{xx}(L,t)\dot{y}_x(L,t) - y_{xx}(0,t)\dot{y}_x(0,t) - y_{xxx}(L,t)\dot{y}(L,t) + y_{xxx}(0,t)\dot{y}(0,t)$$

$$+ \int_0^L y_{xxxx}(x,t)\dot{y}(x,t)dx$$

$$= -y_{xxx}(L,t)\dot{y}(L,t) + \int_0^L y_{xxxx}(x,t)\dot{y}(x,t)dx$$

Using above equation, combing (4.25) and (4.26), we have

$$\dot{V}_f(t) = \rho \int_0^L \dot{y}(x,t)\ddot{y}(x,t)dx + EI \int_0^L y_{xx}(x,t)\dot{y}_{xx}(x,t)dx + m\dot{y}(L,t)\ddot{y}(L,t)$$

$$= -EI \int_0^L \dot{y}(x,t)y_{xxxx}(x,t)dx - EI y_{xxx}(L,t)\dot{y}(L,t)$$

$$+ EI \int_0^L y_{xxxx}(x,t)\dot{y}(x,t)dx + m\dot{y}(L,t)\ddot{y}(L,t)$$

$$= -EI y_{xxx}(L,t)\dot{y}(L,t) + \dot{y}(L,t)(EI y_{xxx}(L,t) + F_f) = \dot{y}(L,t)F_f$$

Applying (4.27) into above yields

$$\dot{V}_f(t) = -k_3 \dot{y}^2(L,t) \leq 0$$

Hence, the operator $\mathcal{A}$ is dissipative.

Second step: Unique Analysis of Solutions

To prove the unique solutions of q and $\mathcal{A}^{-1}$ is a compact operator, define $g = [g_1 \quad g_2 \quad g_3 \quad g_4]^T \in \mathcal{H}$, let

$$\mathcal{A}q = g \qquad (4.33)$$

From $\mathcal{A}q = \begin{bmatrix} q_2 \\ -\frac{EI}{\rho}q_{1,xxxx} \\ q_4 \\ \frac{EI}{m}q_{3,xxx} - \frac{k_3}{m}q_4 \end{bmatrix}$, we have

4.4 Boundary Controller Design

$$g_1 = q_2$$
$$g_2 = -\frac{EI}{\rho} q_{1,xxxx}$$
$$g_3 = q_4$$
$$g_4 = \frac{EI}{m} q_{3,xxx} - \frac{k_3}{m} q_4$$

The solution of $g_2 = -\frac{EI}{\rho} q_{1,xxxx}$ is

$$q_1 = -\frac{\rho}{EI} \int_0^x \int_0^{\xi_1} \int_0^{\xi_2} \int_0^{\xi_3} g_2(\xi_4) d\xi_4 d\xi_3 d\xi_2 d\xi_1 + \sum_{j=0}^{3} \sigma_j x^j$$

where $\sigma_0, \ldots, \sigma_3$ are uniquely determined by boundary conditions (4.17)–(4.20). Then we can get the unique solution as

$$\begin{aligned} q_1 &= -\frac{\rho}{EI} \int_0^x \int_0^{\xi_1} \int_0^{\xi_2} \int_0^{\xi_3} g_2(\xi_4) d\xi_4 d\xi_3 d\xi_2 d\xi_1 + \sum_{j=0}^{3} \sigma_j x^j \\ q_2 &= g_1 \\ q_3 &= q_1(L) \\ q_4 &= g_3 \end{aligned} \quad (4.34)$$

Hence, Equation $\mathcal{A}q = g$ has a unique solution $q \in \mathcal{D}(\mathcal{A})$, implying that $\mathcal{A}^{-1}$ exists and maps $\mathcal{H}$ into $H^4 \times H^2 \times \mathbb{R}^2$. Moreover, since $\mathcal{A}^{-1}$ maps every bounded set of $\mathcal{H}$ into bounded set of $H^4 \times H^2 \times \mathbb{R}^2$, the embedding of the later space onto $\mathcal{H}$ is compact. It follows that $\mathcal{A}^{-1}$ is a compact operator.

The spectrum of $\mathcal{A}$ consists entirely of isolated eigen-values. It also proves that for any $\lambda > 0$ in the resolvent set of $\mathcal{A}$, the operator $(\lambda I - \mathcal{A})^{-1}$ is a compact operator. Based on the Lumer-Phillips theorem, operator $\mathcal{A}$ generates a C_0-semi-group of contractions $T(t)$ on $\mathcal{H}$ [4].

Third step: Convergence Analysis

When we let $\dot{V}_f \equiv 0$, we have $\dot{y}(L,t) \equiv 0$ and $\ddot{y}(L,t) \equiv 0$. from $\ddot{y}(x,t) = -\frac{EI}{\rho} y_{xxxx}(x,t)$, we have $y_{xxxx}(L,t) = 0$.

Note that Eq. (4.25) is separable and can be treated by the technique of separation of variables [5]. We can write $y(x,t)$ as follows:

$$y(x,t) = X(x) \cdot T(t) \quad (4.35)$$

where $X(x)$ and $T(t)$ are unknown functions.

From $\ddot{y}(x,t) = -\frac{EI}{\rho} y_{xxxx}(x,t)$, we have

$$y_{xxxx}(x,t) = -\frac{\rho}{EI} \ddot{y}(x,t)$$

From (4.35), we have $y_{xxxx}(x,t) = X^{(4)}(x) \cdot T(t)$, $\ddot{y}(x,t) = X(x) \cdot T^{(2)}(t)$, then above equation becomes

$$\frac{X^{(4)}(x)}{X(x)} = -\frac{\rho}{EI} \frac{T^{(2)}(t)}{T(t)} = \mu$$

i.e.

$$X^{(4)}(x) - \mu X(x) = 0$$

Let $\mu = \beta^4$, then we can get the solution as

$$X(x) = c_1 \cosh \beta x + c_2 \sinh \beta x + c_3 \cos \beta x + c_4 \sin \beta x \quad (4.36)$$

where $c_i \in R$, $i = 1, 2, 3, 4$ are unknown real number.

Since $y(0,t) = y_x(0,t) = y_{xx}(L,t) = 0$, considering $y_{xxxx}(L,t) = 0$ and $y(x,t) = X(x) \cdot T(t)$, we have $X(0) = X'(0) = X''(L) = X^{(4)}(L) = 0$. Based on (4.36), we can conclude that $X^{(4)}(x) - \mu X(x) = 0$ have unique solutions, $c_i = 0$, $i = 1, 2, 3, 4$, thus, the solutions are $X(x) = 0$, i.e., $y(x,t) = 0$.

Therefore, according to the extended LaSalle's invariance principle [6], the PDE boundary control (4.27) can guarantee the asymptotical stability of the fast subsystem, i.e. for $x \in [0, L]$, when $t \to \infty$, $y(x,t) \to 0$, $\dot{y}(L,t) \to 0$.

4.4.3 Total Boundary Controller

According to the literatures [1, 7, 8], for the singular perturbation system, the stable control law can be designed according to the fast and slow system respectively, and the composite control is stable.

From (4.22), (4.23) and (4.27), we can get the total controller as

$$\tau = \tau_s + \tau_f = -I_h(\dot{c}\dot{e} - \ddot{\theta}_d) - \eta_1 \text{sgn}(s_s) - k_1 s_s \quad (4.37)$$

$$F = F_s + F_f = -\left(M_t L + \frac{1}{2}\rho L^2\right)(\dot{c}\dot{e} - \ddot{\theta}_d) - \eta_2 \text{sgn}(s_s) - k_2 s_s - k_3 s_f \quad (4.38)$$

4.5 Simulation Example

Consider the PDE model as Eqs. (4.1)–(4.5), let $\gamma_1 = \gamma_2 = 0$, choose parameters as: $EI = 25$, $\rho = 0.52$, $m = 2.0$, $L = 1.0$, $I_h = 0.04$, and choose $d_1(t) = \sin t$ and $d_2(t) = \sin t$.

Define ideal angle as $\theta_d = 0.5$, use controller (4.37) and (4.38), let $k_1 = k_2 = k_3 = 50$, $c = 30$, $\eta_1 = 1.5$, $\eta_2 = 1.5$. In (4.22) and (4.23), we use saturation function to replace switch function, and choose the boundary layer thickness as $\Delta = 0.05$. Two axes are divided into sections according to $nx = 10$, $nt = 20000$. The simulation results are shown from Figs. 4.1, 4.2, 4.3, 4.4 and 4.5.

```
%Simulation program: chap4_1.m
close all;
clear all;
nx=10;nt=20000;
tmax=10.0;L=1.0;
%Compute mesh spacing and time step
dx=L/(nx-1);
T=tmax/(nt-1);

%Create arrays to save data for export
t=linspace(0,nt*T,nt);
x=linspace(0,L,nx);
%Parameters
EI=25;m=2.0;rho=0.52;Ih=0.04;
gama1=0;gama2=0;

k1=50;k2=50;k3=50;
c=30;
yL_1=0;
dyLj=0;
yxx0_1=0;
yxxx_L0=0;
%Define variables and Initial condition:
y=zeros(nx,nt);    %elastic deflectgion
th_2=0;th_1=0;
for j=1:nt
    th(j)=0;   %joint angle
end

for j=3:nt    %Begin
thd(j)=0.5;
dthd=0;ddthd=0;
```

```
dt1(j)=1*sin(j*T);
dt2(j)=1*sin(j*T);
%Boundary control input
dth=(th_1-th_2)/T;

e=th_1-thd(j);
de=dth-dthd;

s=c*e+de;

xite1=1.5;
xite2=1.5;

delta=0.05;
kk=1/delta;
if abs(s)>delta
   sats=sign(s);
else
   sats=kk*s;
end

tols=-Ih*(c*de-ddthd)-xite1*sats-k1*s;   %(4.22)
tolf=0;              %(4.27)
tol(j)=tols+tolf;    %(4.37)

Fs=-(m*L+0.5*rho*L*L)*(c*de-ddthd)-xite2*sats-k2*s;  %(4.23)
Ff=-k3*dyLj;         %(4.27)
F(j)=Fs+Ff;          %(4.38)

%th(j)
yxx0=(y(3,j-1)-2*y(2,j-1)+y(1,j-1))/dx^2;

th(j)=2*th_1-th_2+T^2/Ih*(tol(j)+EI*yxx0+dt1(j));           %(4.2)
%%%%%%%%%%%%%%%%%%%%%%%%%
%y(i,j)
ddth(j)=(th(j)-2*th_1+th_2)/T^2;

%get y(i,j),i=1,2, Boundary condition (4.3)
y(1,:)=0;    %y(0,t)=0, i=1
```

4.5 Simulation Example

```
y(2,:)=0;      %y(1,t)=0, i=2

%get y(i,j),i=3:nx-2
for i=3:nx-2

yxxxx=(y(i+2,j-1)-4*y(i+1,j-1)+6*y(i,j-1)-4*y(i-1,j-1)+y(i-2,j-1))/dx
^4;

y(i,j)=T^2*(-i*dx*ddth(j)-EI*yxxxx/rho)+2*y(i,j-1)-y(i,j-2);   %i*dx=x
,(4.1)
end

%get y(nx-1,j),i=nx-1
yxxxx(nx-1,j-1)=(-2*y(nx,j-1)+5*y(nx-1,j-1)-4*y(nx-2,j-1)+y(nx-3,j-1)
)/dx^4;
y(nx-1,j)=T^2*(-(nx-1)*dx*ddth(j)-EI*yxxxx(nx-1,j-1)/rho)+2*y(nx-1,j-
1)-y(nx-1,j-2);   %(4.1)

%get y(nx,j),y=nx
yxxx_L=(-y(nx,j-1)+2*y(nx-1,j-1)-y(nx-2,j-1))/dx^3;
y(nx,j)=T^2*(-L*ddth(j)+(EI*yxxx_L+F(j)+dt2(j))/m)+2*y(nx,j-1)-y(nx,j
-2) ;    %(4.5)

dyL(j)=(y(nx,j-1)-y(nx,j-2))/T;
yL_1=y(nx,j);
%%%%%%%%%%%%%%%%%%%%%%%%%
th_2=th_1;
th_1=th(j);
yxx0_1=yxx0;
yxxx_L0=yxxx_L;
dyLj=dyL(j);
end

%To view the curve, short the points
tshort=linspace(0,tmax,nt/100);
yshort=zeros(nx,nt/100);
for j=1:nt/100
    for i=1:nx
        yshort(i,j)=y(i,j*100);    %Using true y(i,j)
        %dyL(j)=dyL(j*100);
    end
end
%%%%%%%%%%%%%%%%%%%%%%%%%
figure(1);
```

```
plot(t,thd,'k',t,th,'r','linewidth',2);
legend('Ideal angle signal','Angle response');
xlabel('time');ylabel('theta');

figure(2);
surf(tshort,x,yshort);
xlabel('time'); ylabel('x');zlabel('Deflection,y(x,t)');

figure(3);
subplot(2,1,1)
plot(t,tol,'k','linewidth',2);
xlabel('time');ylabel('Control input£¬tol');
subplot(2,1,2)
plot(t,F,'k','linewidth',2);
xlabel('time');ylabel('Control input,F');

figure(4);
subplot(211);
for j=1:nt/100
    yshortL(j)=y(nx,j*100);
end
plot(tshort,yshortL,'k','linewidth',2);
xlabel('time');ylabel('y(L,t)');
subplot(212);
for j=1:nt/100
    yshort1(j)=y(nx/2,j*100);
end
plot(tshort,yshort1,'k','linewidth',2);
xlabel('time');ylabel('y(x,t) at half of L');

figure(5);
plot(t,dyL,'k','linewidth',2);
xlabel('time');ylabel('dy(L,t)');
```

4.5 Simulation Example

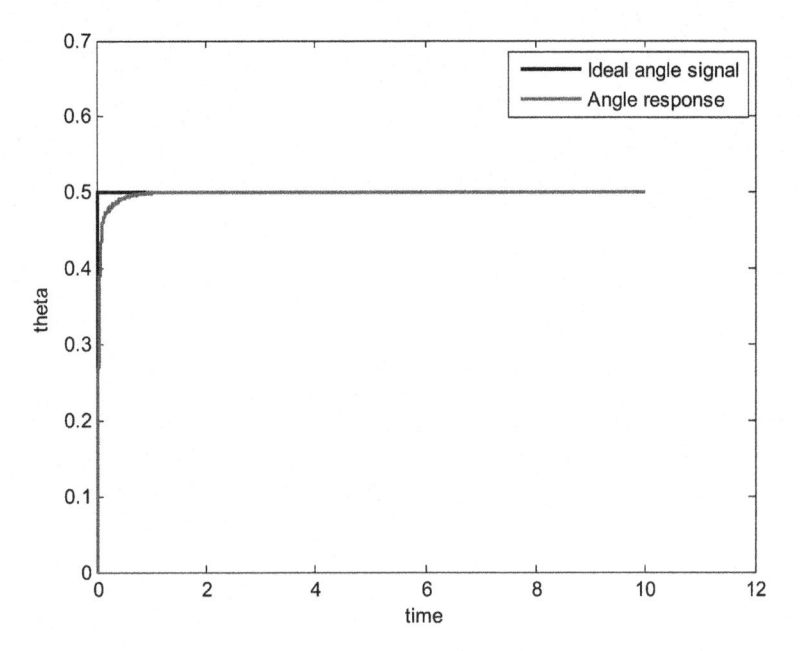

Fig. 4.1 Angle response

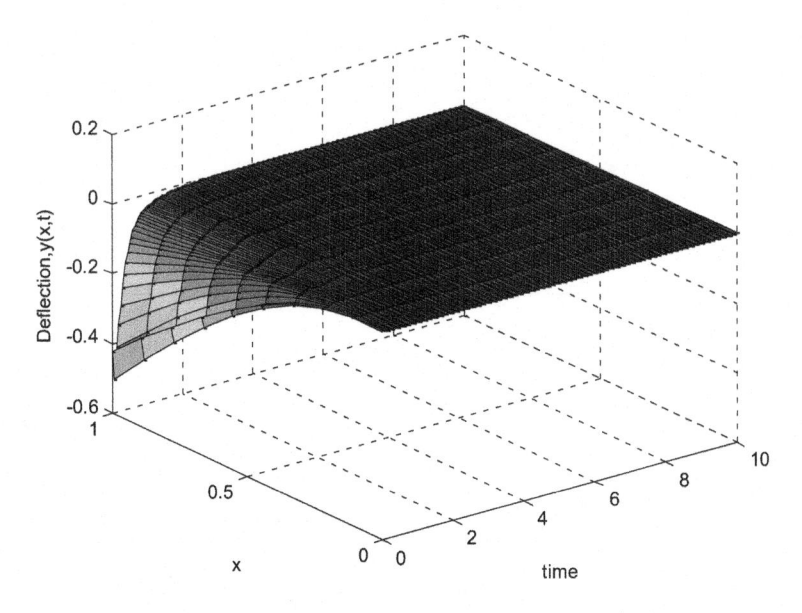

Fig. 4.2 The distributed elastic deflection of the flexible manipulator, $y(x,t) \to 0$

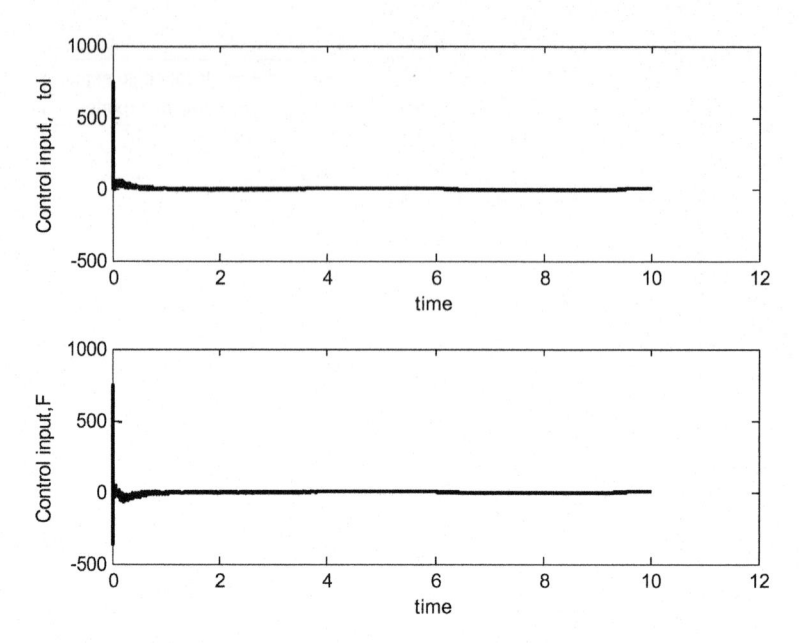

Fig. 4.3 Control input of τ and F

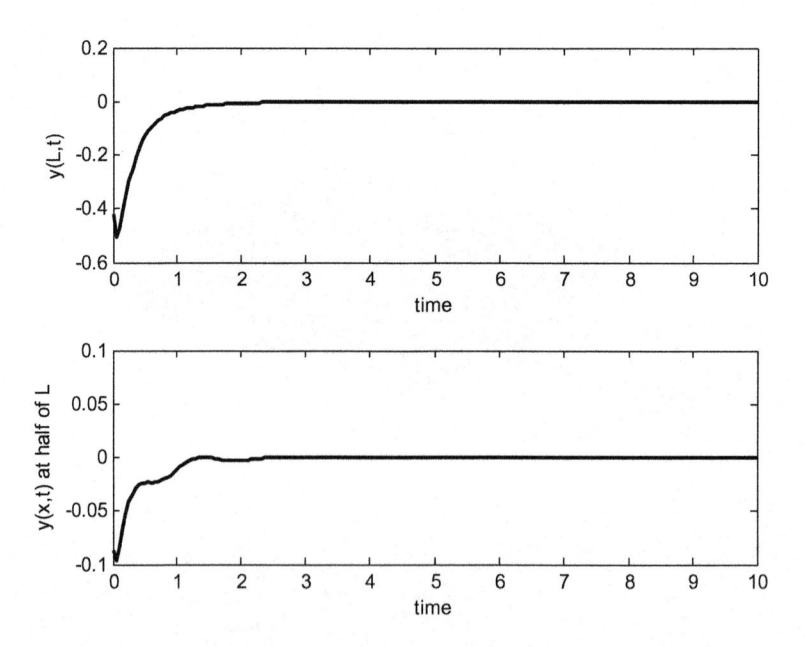

Fig. 4.4 The elastic deflection of the flexible manipulator at $x = 0.5L$ and $x = L$

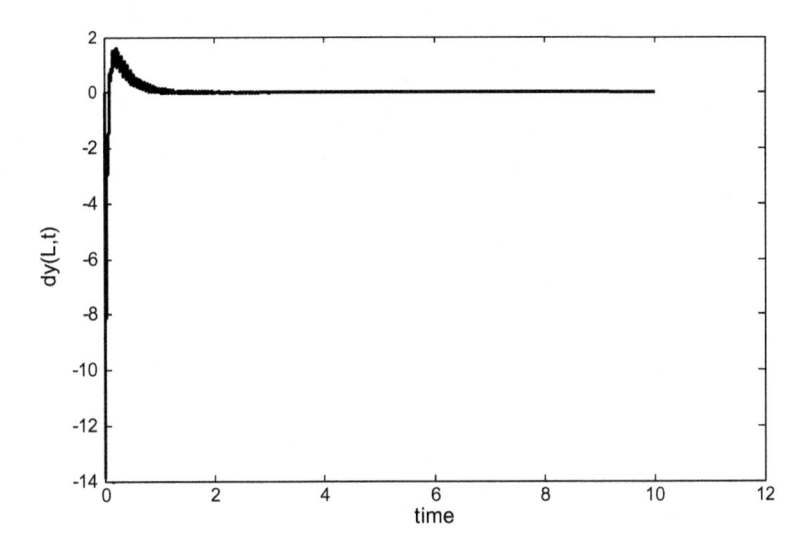

Fig. 4.5 The elastic deflection rate of the flexible manipulator at $x = L$, $\dot{y}(L,t) \to 0$

References

1. P.V. Kokotovic, Applications of singular perturbation techniques to control problems. SIAM Rev. **26**(4), 501–550 (1984)
2. B. Siciliano, W.J. Book, A singular perturbation approach to control of lightweight flexible manipulators. Int. J. Robot. Res. **7**(4), 79–90 (1988)
3. Y. Aoustin, C. Chevallereau, The singular perturbation control of a two-flexible-link robot, in *IEEE International Conference on Robotics and Automation*, 1993, pp. 737–742
4. Z.J. Liu, J.K. Liu, Adaptive boundary control of a flexible manipulator with input saturation. Int. J. Control **89**(6), 1191–1202 (2016)
5. W.H. Ray, *Advanced Process Control* (McGraw-Hill, New York, NY, USA, 1981)
6. C.D. Rahn, *Mechatronic Control of Distributed Noise and Vibration-A Lyapunov Approach* (Springer, Berlin Heidelberg, 2001)
7. H.K. Khalil, Output feedback control of linear two-time-scale systems. IEEE Trans. Autom. Control **32**(9), 784–792 (1987)
8. A. Saberi, H. Khalil, Quadratic-type Lyapunov functions for singularly perturbed systems. IEEE Trans. Autom. Control **29**(6), 542–550 (1984)

Chapter 5
Boundary Control for Flexible Manipulator with Exponential Convergence

The flexible manipulator system will cause small deformation and elastic vibration because of its flexible characteristics. It will become large deformation vibration if the flexibility is large enough. The motion and vibration of a flexible manipulator are coupled and interacting with each other, which will interfere with the performance of the robot arm. In severe cases, the elastic vibration of the boom will destroy the stability of the whole system and even make the whole robot system out of control and invalid. Therefore, how to weaken the vibration of the flexible manipulator in motion is an urgent problem that needs to be solved.

In this chapter, to weaken the vibration of the manipulator, the boundary control method with exponential convergence is designed.

5.1 System Description

Considering PDE model given in Chap. 3, the PDE model is given as

$$\rho \ddot{z}(x) = -EI z_{xxxx}(x) \tag{5.1}$$

$$\tau = I_h \ddot{z}_x(0) - EI z_{xx}(0) \tag{5.2}$$

$$F = m\ddot{z}(L) - EI z_{xxx}(L) \tag{5.3}$$

$$z(0) = 0, \; z_x(0) = \theta, \; z_{xx}(L) = 0 \tag{5.4}$$

where $z(x) = x\theta + y(x)$, $\ddot{z}(x) = x\ddot{\theta} + \ddot{y}(x)$, $\ddot{z}(L) = L\ddot{\theta} + \ddot{y}(L)$.
From $z(x)$ definition, we have

$$z_{xx}(x) = y_{xx}(x), \ddot{z}_x(0) = \ddot{\theta}, z_{xx}(0) = y_{xx}(0), z_{xx}(L) = y_{xx}(L), z_{xxx}(L) = y_{xxx}(L) \tag{5.5}$$

The control goal is: $\theta(t) \to \theta_d(t)$, $\dot{\theta}(t) \to \dot{\theta}_d(t)$, $y(x,t) \to 0$, $\dot{y}(x,t) \to 0$, $\theta_d(t)$ is ideal angle signal.

5.2 Some Lemmas

Lemma 5.1 ([1]) For $\phi_1(x,t), \phi_2(x,t) \in R$, $x \in [0,L]$, $t \in [0,\infty)$, there exists

$$\begin{aligned}\phi_1(x,t)\phi_2(x,t) &\leq |\phi_1(x,t)\phi_2(x,t)| \leq \phi_1^2(x,t) + \phi_2^2(x,t) \\ |\phi_1(x,t)\phi_2(x,t)| &\leq \tfrac{1}{\gamma}\phi_1^2(x,t) + \gamma\phi_2^2(x,t)\end{aligned} \quad (5.6)$$

where $\gamma > 0$.

Lemma 5.2 ([2]) For $p(x,t) \in R$, $x \in [0,L]$, $t \in [0,\infty)$, if $p(0,t) = 0, \forall t \in [0,\infty)$, then

$$p^2(x,t) \leq L \int_0^L p_x^2(x,t)dx, \forall x \in [0,L] \quad (5.7)$$

similarly, if $p_x(0,t) = 0, \forall t \in [0,\infty)$, then

$$p_x^2(x,t) \leq L \int_0^L p_{xx}^2(x,t)dx, \forall x \in [0,L] \quad (5.8)$$

Lemma 5.3 ([3]) For $V : [0,\infty) \in R$, $\forall t \geq t_0 \geq 0$, if $\dot{V} \leq -\eta V + g$, then

$$V(t) \leq e^{-\eta(t-t_0)}V(t_0) + \int_{t_0}^t e^{-\eta(t-s)}g(s)ds \quad (5.9)$$

where $\eta > 0$.

5.3 Boundary Controller Design

To realize angle response and weaken the vibration of the manipulator, design boundary controller as [4]

$$\tau = -k_p e - k_d \dot{e} \quad (5.10)$$

5.3 Boundary Controller Design

$$F = -ku_a + m\dot{z}_{xxx}(L) \tag{5.11}$$

where $k_p > 0$, $k_d > 0$, $k > 0$, $u_a = \dot{z}(L) - z_{xxx}(L)$, $e = \theta - \theta_d$, θ_d is ideal angle with constant value, $\dot{e} = \dot{\theta} - \dot{\theta}_d = \dot{\theta}$, $\ddot{e} = \ddot{\theta} - \ddot{\theta}_d = \ddot{\theta}$.

Theorem 5.1 ([4]) *Using controller* (5.10) *and* (5.11), *the closed system will be stable, when* $t \to \infty$, *for* $x \in [0, L]$, $\theta \to \theta_d$, $\dot{\theta} \to 0$, $y(x) \to 0$, $\dot{y}(x) \to 0$ *exponentially*.

Proof *Design* Lyapunov function as

$$V(t) = E_1 + E_2 + E_a \tag{5.12}$$

where

$$E_1 = \frac{1}{2}\int_0^L \rho \dot{z}^2(x)dx + \frac{1}{2}EI\int_0^L y_{xx}^2(x)dx \tag{5.13}$$

$$E_2 = \frac{1}{2}I_h \dot{e}^2 + \frac{1}{2}k_p e^2 + \frac{1}{2}mu_a^2 \tag{5.14}$$

$$E_a = \alpha\rho \int_0^L x\dot{z}(x)ze_x(x)dx + \alpha I_h e\dot{e} \tag{5.15}$$

where E_1 is sum of the kinetic energy and potential energy of the manipulator, which indicates the index of bending deformation and bending deformation rate of the manipulator. The first two items of E_2 represent angle tracking error index, the third item is the auxiliary term. E_a is the auxiliary term, α is a small positive real number, and

$$ze(x) = xe + y(x), ze_x(x) = e + y_x(x), ze_{xx} = y_{xx}(x) = z_{xx}(x) \tag{5.16}$$

From Lemma 5.1, we have

$$x\dot{z}(x)ze_x(x) \leq |x\dot{z}(x)ze_x(x)| \leq L(\dot{z}^2(x) + ze_x^2(x))$$

Since $y(0) = 0$, from Lemma 5.2 and $z_{xx}(x) = y_{xx}(x)$, we have

$$2\alpha\rho L\int_0^L y_x^2(x)dx \leq 2\alpha\rho L\int_0^L L\int_0^L y_{xx}^2(x,t)dxdx = 2\alpha\rho L^3 \int_0^L z_{xx}^2(x)dx$$

then

$$|E_a| \leq \alpha\rho L \int_0^L \left(\dot{z}^2(x) + ze_x^2(x)\right)dx + \alpha I_h\left(e^2 + \dot{e}^2\right)$$

$$= \alpha\rho L \int_0^L \left(\dot{z}^2(x) + e^2 + y_x^2(x) + 2e \cdot y_x(x)\right)dx + \alpha I_h\left(e^2 + \dot{e}^2\right)$$

$$\leq \alpha\rho L \int_0^L \left(\dot{z}^2(x) + 2e^2 + 2y_x^2(x)\right)dx + \alpha I_h\left(e^2 + \dot{e}^2\right)$$

$$= \alpha\rho L \int_0^L \dot{z}^2(x)dx + 2\alpha\rho L^2 e^2 + 2\alpha\rho L \int_0^L y_x^2(x)dx + \alpha I_h\left(e^2 + \dot{e}^2\right) \quad (5.17)$$

$$\leq \alpha\rho L \int_0^L \dot{z}^2(x)dx + 2\alpha\rho L^2 e^2 + 2\alpha\rho L^3 \int_0^L z_{xx}^2(x)dx + \alpha I_h\left(e^2 + \dot{e}^2\right)$$

$$= \alpha\rho L \int_0^L \dot{z}^2(x)dx + 2\alpha\rho L^3 \int_0^L z_{xx}^2(x)dx + \left(\alpha I_h + 2\alpha\rho L^2\right)e^2 + \alpha I_h \dot{e}^2$$

$$\leq \alpha_1(E_1 + E_2)$$

where $\alpha_1 = \max\left(2\alpha L, \frac{4\alpha\rho L^3}{EI}, \frac{2(\alpha I_h + 2\alpha\rho L^2)}{k_p}, 2\alpha\right)$.

then

$$-\alpha_1(E_1 + E_2) \leq E_a \leq \alpha_1(E_1 + E_2) \quad (5.18)$$

Set $0 < \alpha_1 < 1$, i.e. $0 < \max\left(2\alpha L, \frac{2\alpha\rho L^3}{EI}, \frac{2(\alpha I_h + 2\alpha\rho L^2)}{k_p}, 2\alpha\right) < 1$, then α can be designed as

$$0 < \alpha < \frac{1}{\max\left(2L, \frac{2\rho L^3}{EI}, \frac{2(I_h + 2\rho L^2)}{k_p}, 2\right)} \quad (5.19)$$

Define $1 > \alpha_2 = 1 - \alpha_1 > 0$, $2 > \alpha_3 = 1 + \alpha_1 > 1$, then

$$0 \leq \alpha_2(E_1 + E_2) \leq E_a + E_1 + E_2 \leq \alpha_3(E_1 + E_2)$$

and

$$0 \leq \alpha_2(E_1 + E_2) \leq V(t) \leq \alpha_3(E_1 + E_2) \quad (5.20)$$

5.3 Boundary Controller Design

From (5.20), we can guarantee Lyapunov function $V(t)$ is a positive definite function, and

$$\dot{V}(t) = \dot{E}_1 + \dot{E}_2 + \dot{E}_a \tag{5.21}$$

where

$$\dot{E}_1 = \int_0^L \rho \dot{z}(x) \ddot{z}(x) dx + EI \int_0^L y_{xx}(x) \dot{y}_{xx}(x) dx \tag{5.22}$$

$$\dot{E}_2 = I_h \dot{e} \ddot{e} + k_p e \dot{e} + m u_a \dot{u}_a \tag{5.23}$$

$$\dot{E}_a = \dot{E}_{a1} + \dot{E}_{a2} + \dot{E}_{a3} \tag{5.24}$$

$$\dot{E}_{a1} = \alpha \rho \int_0^L x \ddot{z}(x) z e_x(x) dx \tag{5.25}$$

$$\dot{E}_{a2} = \alpha \rho \int_0^L x \dot{z}(x) \dot{z} e_x(x) dx \tag{5.26}$$

$$\dot{E}_{a3} = \alpha I_h \left(\dot{e}^2 + e \ddot{e} \right) \tag{5.27}$$

Submitting (5.1), i.e. $\rho \ddot{z}(x) = -EI z_{xxxx}(x)$ into (5.22), we have

$$\dot{E}_1 = -EI \int_0^L \dot{z}(x) z_{xxxx}(x) dx + EI \int_0^L y_{xx}(x) \dot{y}_{xx}(x) dx$$

$$\int_0^L \dot{z}(x) z_{xxxx}(x) dx = \int_0^L \dot{z}(x) d z_{xxx}(x)$$

$$= \dot{z}(x) z_{xxx}(x) \big|_0^L - \int_0^L z_{xxx}(x) \dot{z}_x(x) dx = \dot{z}(L) z_{xxx}(L) - \int_0^L z_{xxx}(x) \dot{z}_x(x) dx$$

$$\int_0^L y_{xx}(x)\dot{y}_{xx}(x)dx = \int_0^L z_{xx}(x)\dot{z}_{xx}(x)dx = \int_0^L z_{xx}(x)d\dot{z}_x(x)$$

$$= z_{xx}(x)\dot{z}_x(x)\big|_0^L - \int_0^L \dot{z}_x(x)z_{xxx}(x)dx = -z_{xx}(0)\dot{\theta} - \int_0^L \dot{z}_x(x)z_{xxx}(x)dx$$

where $z_{xx}(L) = 0$, $\dot{z}_x(0) = \dot{\theta}$.

then

$$\dot{E}_1 = -EI\int_0^L \dot{z}(x)z_{xxxx}(x)dx + EI\int_0^L y_{xx}(x)\dot{y}_{xx}(x)dx$$

$$= -EI\left(\dot{z}(L)z_{xxx}(L) - \int_0^L z_{xxx}(x)\dot{z}_x(x)dx\right) + EI\left(-z_{xx}(0)\dot{\theta} - \int_0^L \dot{z}_x(x)z_{xxx}(x)dx\right)$$

$$= -EI\dot{z}(L)y_{xxx}(L) - EIy_{xx}(0)\dot{\theta}$$

i.e.

$$\dot{E}_1 = -EIy_{xxx}(L)\dot{z}(L) - EIy_{xx}(0)\dot{\theta} \tag{5.28}$$

From (5.4)–(5.5) and $u_a = \dot{z}(L) - z_{xxx}(L)$, (5.28) becomes

$$\dot{E}_1 = -EIz_{xxx}(L)\dot{z}(L) - EIz_{xx}(0)\dot{e} \\
= -EIz_{xx}(0)\dot{e} - EIz_{xxx}^2(L) - EIz_{xxx}(L)u_a \tag{5.29}$$

Considering (5.2) and (5.3), and combining with (5.23) and (5.29), we have

$$\dot{E}_1 + \dot{E}_2 = -EIz_{xx}(0)\dot{e} - EIz_{xxx}^2(L) - EIz_{xxx}(L)u_a + \dot{e}(I_h\ddot{e} + k_p e) + mu_a\dot{u}_a \\
= \dot{e}(I_h\ddot{e} + k_p e - EIz_{xx}(0)) - EIz_{xxx}^2(L) + u_a(-EIy_{xxx}(L) + m\dot{u}_a) \tag{5.30} \\
= \dot{e}(\tau + k_p e) + u_a(F - m\ddot{z}_{xxx}(L)) - EIz_{xxx}^2(L)$$

Submitting (5.10) and (5.11) into above, we have

$$\dot{E}_1 + \dot{E}_2 = -k_d\dot{e}^2 - ku_a^2 - EIz_{xxx}^2(L) \tag{5.31}$$

Submitting (5.1) into (5.25), we have

$$\dot{E}_{a1} = \alpha\int_0^L x(-EIz_{xxxx}(x))ze_x(x)dx = -\alpha EI\int_0^L xz_{xxxx}(x)ze_x(x)dx \tag{5.32}$$

5.3 Boundary Controller Design

Using partial integration for (5.32), we have

$$\int_0^L xz_{xxxx}(x)ze_x(x)dx = \int_0^L xze_x(x)dz_{xxx}(x) = xze_x(x)\cdot z_{xxx}(x)\Big|_0^L - \int_0^L z_{xxx}(x)d(xze_x(x))$$

$$= Lze_x(L)\cdot z_{xxx}(L) - \int_0^L z_{xxx}(x)(ze_x(x)+xze_{xx}(x))dx$$

$$= Lze_x(L)\cdot z_{xxx}(L) - \int_0^L z_{xxx}(x)ze_x(x)dx - \int_0^L z_{xxx}(x)xze_{xx}(x)dx$$

$$= A - B - C$$

where $A = Lze_x(L)\cdot z_{xxx}(L)$, $B = \int_0^L z_{xxx}(x)ze_x(x)dx$, $C = \int_0^L z_{xxx}(x)xze_{xx}(x)dx$.

Using partial integral method, we have

$$B = \int_0^L z_{xxx}(x)ze_x(x)dx = \int_0^L ze_x(x)dz_{xx}(x)$$

$$= ze_x(x)z_{xx}(x)\Big|_0^L - \int_0^L ze_{xx}(x)z_{xx}(x)dx$$

$$= -ez_{xx}(0) - \int_0^L z_{xx}^2(x)dx$$

$$C = \int_0^L z_{xxx}(x)xze_{xx}(x)dx = xze_{xx}(x)z_{xx}(x)\Big|_0^L - \int_0^L z_{xx}(x)d(xze_{xx}(x))$$

$$= -\int_0^L z_{xx}(x)(ze_{xx}(x)+xze_{xxx}(x))dx$$

$$= -\int_0^L z_{xx}^2(x)dx - \int_0^L z_{xx}(x)xze_{xxx}(x)dx = -\int_0^L z_{xx}^2(x)dx - C$$

i.e., $C = -\frac{1}{2}\int_0^L z_{xx}^2(x)dx$

From above, we have

$$\int_0^L xz_{xxxx}(x)ze_x(x)dx = A - B - C = Lze_x(L)z_{xxx}(L) + \frac{3}{2}\int_0^L z_{xx}^2(x)dx + ez_{xx}(0)$$

then

$$\dot{E}_{a1} = -\alpha EI(A - B - C) = -\alpha EILze_x(L)z_{xxx}(L) - \frac{3}{2}\alpha EI \int_0^L z_{xx}^2(x)dx - \alpha EIez_{xx}(0) \tag{5.33}$$

Submitting (5.6)–(5.8) and (5.16) into (5.33), we have

$$\dot{E}_{a1} \leq \alpha EILze_x^2(L) + \alpha EILz_{xxx}^2(L) - \frac{3}{2}\alpha EI \int_0^L z_{xx}^2(x)dx - \alpha EIez_{xx}(0) + \alpha L \int_0^L ze_x^2(x)dx$$

$$= \alpha EILze_x^2(L) + \alpha EILz_{xxx}^2(L) - \frac{3}{2}\alpha EI \int_0^L z_{xx}^2(x)dx - \alpha EIez_{xx}(0) + \alpha L \int_0^L \left(e^2 + y_x^2(x) + 2e \cdot y_x(x)\right)dx$$

$$\leq \alpha EIL\left(2e^2 + 2L\int_0^L z_{xx}^2(x,t)dx\right) + \alpha EILz_{xxx}^2(L) - \frac{3}{2}\alpha EI \int_0^L z_{xx}^2(x)dx - \alpha EIez_{xx}(0)$$

$$+ 2\alpha e^2 L^2 + 2\alpha L^3 \int_0^L z_{xx}^2(x,t)dx$$

$$\leq -\left(\frac{3}{2}\alpha - 2\alpha L^2 - \frac{2\alpha L^3}{EI}\right)\int_0^L EIz_{xx}^2(x)dx + \alpha EILz_{xxx}^2(L) - \alpha EIez_{xx}(0) + (2\alpha EIL + 2\alpha L^2)e^2 \tag{5.34}$$

From $y_x(0) = 0$, we have $y_x^2(x) \leq L\int_0^L z_{xx}^2(x,t)dx, \forall x \in [0, L]$ and $y_x^2(L) \leq L\int_0^L z_{xx}^2(x,t)dx$, then $\int_0^L\int_0^L z_{xx}^2(x,t)dxdx = L\int_0^L z_{xx}^2(x,t)dx$, and

$$\alpha EILze_x^2(L) = \alpha EIL\left(e^2 + y_x^2(L) + 2e \cdot y_x(L)\right)$$

$$\leq \alpha EIL\left(2e^2 + 2y_x^2(L)\right) \leq \alpha EIL\left(2e^2 + 2L\int_0^L z_{xx}^2(x,t)dx\right)$$

5.3 Boundary Controller Design

$$\alpha L \int_0^L \left(e^2 + y_x^2(x) + 2e \cdot y_x(x)\right) dx \leq \alpha L \int_0^L \left(2e^2 + 2y_x^2(x)\right) dx$$

$$\leq \alpha L \int_0^L \left(2e^2 + 2L \int_0^L z_{xx}^2(x,t) dx\right) dx$$

$$\leq 2\alpha e^2 L^2 + 2\alpha L^2 \int_0^L \int_0^L z_{xx}^2(x,t) dx dx$$

$$\leq 2\alpha e^2 L^2 + 2\alpha L^3 \int_0^L z_{xx}^2(x,t) dx$$

$$\alpha L \int_0^L z e_x^2(x) dx = \alpha L \int_0^L (e + y_x(x))^2 dx = \alpha L \int_0^L \left(e^2 + y_x^2(x) + 2e \cdot y_x(x)\right) dx$$

From (5.26), using partial integral method, we have

$$\dot{E}_{a2} = \frac{1}{2}\alpha \rho L \dot{z}^2(L) - \frac{1}{2}\alpha \rho \int_0^L \dot{z}^2(x) dx \qquad (5.35)$$

From (5.2), we get $\tau = I_h \ddot{\theta} - EI z_{xx}(0) = I_h \ddot{e} - EI z_{xx}(0)$, then $I_h \ddot{e} - EI z_{xx}(0) = -k_p e - k_d \dot{e}$, i.e., $I_h \ddot{e} = EI z_{xx}(0) - k_p e - k_d \dot{e}$, according to Lemma 5.1, we have $-e\dot{e} \leq e^2 + \dot{e}^2$, then

$$\begin{aligned}\dot{E}_{a3} &= \alpha I_h \dot{e}^2 + \alpha I_h e \ddot{e} \\ &= \alpha I_h \dot{e}^2 + \alpha e EI z_{xx}(0) - \alpha k_p e^2 - k_d \alpha e \dot{e} \\ &\leq (\alpha I_h + k_d \alpha)\dot{e}^2 - (\alpha k_p - k_d \alpha)e^2 + \alpha e EI z_{xx}(0)\end{aligned} \qquad (5.36)$$

From (5.34)–(5.36), we have

$$\dot{E}_a = \dot{E}_{a1} + \dot{E}_{a2} + \dot{E}_{a3}$$

$$\leq -\left(\frac{3}{2}\alpha - 2\alpha L^2 - \frac{2\alpha L^3}{EI}\right)\int_0^L EIz_{xx}^2(x)dx + \alpha EILz_{xxx}^2(L) + (2\alpha EIL + 2\alpha L^2)e^2$$

$$+ \frac{1}{2}\alpha\rho L\dot{z}^2(L) - \frac{1}{2}\alpha\rho\int_0^L \dot{z}^2(x)dx + (\alpha I_h + k_d\alpha)\dot{e}^2 - (\alpha k_p - k_d\alpha)e^2$$

$$= -\left(\frac{3}{2}\alpha - 2\alpha L^2 - \frac{2\alpha L^3}{EI}\right)\int_0^L EIz_{xx}^2(x)dx + \alpha EILz_{xxx}^2(L)$$

$$+ \frac{1}{2}\alpha\rho L\dot{z}^2(L) - \frac{1}{2}\alpha\rho\int_0^L \dot{z}^2(x)dx + (\alpha I_h + k_d\alpha)\dot{e}^2 - (\alpha k_p - k_d\alpha - 2\alpha EIL - 2\alpha L^2)e^2$$

(5.37)

then

$$\dot{V}(t) = \dot{E}_1 + \dot{E}_2 + \dot{E}_a$$

$$\leq -k_d\dot{e}^2 - ku_a^2 - EIz_{xxx}^2(L) - \left(\frac{3}{2}\alpha - 2\alpha L^2 - \frac{2\alpha L^3}{EI}\right)\int_0^L EIz_{xx}^2(x)dx + \alpha EILz_{xxx}^2(L)$$

$$+ \frac{1}{2}\alpha\rho L\dot{z}^2(L) - \frac{1}{2}\alpha\rho\int_0^L \dot{z}^2(x)dx + (\alpha I_h + k_d\alpha)\dot{e}^2 - (\alpha k_p - k_d\alpha - 2\alpha EIL - 2\alpha L^2)e^2$$

$$= -\left(\frac{3}{2}\alpha - 2\alpha L^2 - \frac{2\alpha L^3}{EI}\right)\int_0^L EIz_{xx}^2(x)dx - \frac{1}{2}\alpha\int_0^L \rho\dot{z}^2(x)dx - (k_d - \alpha I_h - k_d\alpha)\dot{e}^2$$

$$- (\alpha k_p - k_d\alpha - 2\alpha EIL - 2\alpha L^2)e^2 - ku_a^2 + \frac{1}{2}\alpha\rho L\dot{z}^2(L) - (EI - \alpha EIL)z_{xxx}^2(L)$$

(5.38)

By choosing α, we can guarantee $EI - \alpha EIL > \frac{1}{2}\alpha\rho L$, and

$$\frac{1}{2}\alpha\rho L\dot{z}^2(L) - (EI - \alpha EIL)z_{xxx}^2(L) \leq \eta_0(\dot{z}(L) - z_{xxx}(L))^2 = \eta_0 u_a^2 \quad (5.39)$$

where $\eta_0 > \max\left(\eta_1, \frac{\eta_1\eta_2}{\eta_2 - \eta_1}\right)$.

5.3 Boundary Controller Design

Define $\eta_1 = \frac{1}{2}\alpha\rho L$, $\eta_2 = EI - \alpha EIL$, $a = \dot{z}(L)$, $b = z_{xxx}(L)$, η_0 which must be satisfied as

$$\eta_1 a^2 - \eta_2 b^2 \leq \eta_0 (a-b)^2$$

i.e.

$$(\eta_0 - \eta_1)a^2 - 2\eta_0 ab + (\eta_2 + \eta_0)b^2 \geq 0$$

then

$$(\eta_0 - \eta_1)\left[a^2 - 2\frac{\eta_0}{\eta_0 - \eta_1}ab + \left(\frac{\eta_0}{\eta_0 - \eta_1}\right)^2 b^2\right] + \left((\eta_2 + \eta_0) - (\eta_0 - \eta_1)\left(\frac{\eta_0}{\eta_0 - \eta_1}\right)^2\right)b^2 \geq 0$$

and

$$(\eta_0 - \eta_1)\left(a - \frac{\eta_0}{\eta_0 - \eta_1}b\right)^2 + \left((\eta_2 + \eta_0) - (\eta_0 - \eta_1)\left(\frac{\eta_0}{\eta_0 - \eta_1}\right)^2\right)b^2 \geq 0$$

To ensure the above equations, the following conditions need to be satisfied

$$\begin{cases} \eta_0 - \eta_1 > 0 \\ (\eta_2 + \eta_0) - (\eta_0 - \eta_1)\left(\frac{\eta_0}{\eta_0 - \eta_1}\right)^2 \geq 0 \end{cases}$$

then $(\eta_2 + \eta_0) \geq \frac{\eta_0^2}{\eta_0 - \eta_1}$, $\eta_2\eta_0 - \eta_2\eta_1 - \eta_0\eta_1 \geq 0$. Since $\eta_2 - \eta_1 > 0$, then

$$\eta_0 \geq \frac{\eta_2 \eta_1}{\eta_2 - \eta_1}$$

In summary, η_0 must be satisfied as $\eta_0 > \eta_1$, $\eta_0 \geq \frac{\eta_2\eta_1}{\eta_2-\eta_1}$, i.e. $\eta_0 > \max\left(\eta_1, \frac{\eta_1\eta_2}{\eta_2-\eta_1}\right)$.

According to (5.38) and (5.39), we have

$$\dot{V}(t) \leq -\left(\frac{3}{2}\alpha - 2\alpha L^2 - \frac{2\alpha L^3}{EI}\right)\int_0^L EIz_{xx}^2(x)dx - \frac{1}{2}\alpha\int_0^L \rho\dot{z}^2(x)dx - (k_d - \alpha I_h - k_d\alpha)\dot{e}^2$$
$$- (\alpha k_p - k_d\alpha - 2\alpha EIL - 2\alpha L^2)e^2 - (k - \eta_0)u_a^2$$
$$\leq -\lambda_0(E_1 + E_2) \leq -\lambda_0 \frac{V(t)}{\alpha_3} = -\lambda V(t)$$

To ensure the above equation, the following conditions need to be satisfied

$$\sigma_1 = \frac{3}{2}\alpha - 2\alpha L^2 - \frac{2\alpha L^3}{EI} > 0, \sigma_2 = \frac{1}{2}\alpha > 0, \sigma_3 = k_d - \alpha I_h - k_d\alpha > 0,$$
$$\sigma_4 = \alpha k_p - k_d\alpha - 2\alpha EIL - 2\alpha L^2 > 0 \quad (5.40)$$
$$\sigma_5 = k - \eta_0 > 0, \min\left(2\sigma_1, 2\sigma_2, \frac{2\sigma_3}{I_h}, \frac{2\sigma_4}{k_p}, \frac{2\sigma_5}{m}\right) \geq \lambda_0 > 0, \lambda = \frac{\lambda_0}{\alpha_3}$$

Using Lemma 5.3, the solution of $\dot{V}(t) \leq -\lambda V(t)$ can be solved as

$$V(t) \leq V(0)e^{-\lambda t} \quad (5.41)$$

Since $V(0)$ is limited, we have $t \to \infty$, $V(t) \to 0$ exponentially. From (5.20), we have $E_1 + E_2 \to 0$, then $e \to 0$ and $\dot{e} \to 0$, i.e., $\theta \to \theta_d$, $\dot{\theta} \to 0$ and $\dot{z}(x) \to 0$, from $z(x) = x\theta + y(x)$, we get $\dot{y}(x) \to 0$.

From $E_1 + E_2 \to 0$, we have $\int_0^L y_{xx}^2(x)dx \to 0$, considering $y(0) = y_x(0) = 0$ and Lemma 5.2, we have $y_x^2(x) \leq L\int_0^L y_{xx}^2(x)dx$ and $y^2(x) \leq L\int_0^L y_x^2(x)dx$, then $y(x) \to 0$.

The shortcoming in this chapter is that $z_{xxx}(L)$ and $\dot{z}_{xxx}(L)$ are needed in the controller design, which is difficult in practical engineering.

Remark: To guarantee $\sigma_1 = \frac{3}{2}\alpha - 2\alpha L^2 - \frac{2\alpha L^3}{EI} > 0$, L can not be designed too big value.

5.4 Simulation Example

Consider the PDE model as Eqs. (5.1)–(5.4), the sampling time is set as $\Delta t = 5 \times 10^{-4}$, and set $\Delta x = 0.01$, the physical parameters are chosen as: $EI = 3.0$, $L = 1.0$, $\rho = 0.20$, $m = 0.10$, $I_h = 0.10$.

Define ideal angle as $\theta_d = 0.5$, use controller (5.10) and (5.11), let $k_1 = k_2 = k_3 = 50$, $c = 30$, $\eta_1 = 1.5$, $\eta_2 = 1.5$. Two axes are divided into sections according to $nx = 10$, $nt = 20000$. The simulation results are shown from Figs. 5.1, 5.2 to 5.3.

5.4 Simulation Example

Simulation Programs:

(1) Program for parameter setting: chap5_1.m

```
close all;
clear all;
clc;
%Parameters
EI=3;
rho=0.2;
m=0.1;
Ih=0.1;
L=1.0;

% step (1)
kp=50;

% step (2)
P1=[2*L,2*rho*L^3/EI,2*(Ih+2*rho*L^3)/kp,2];
alfa_max=1/max(P1)
alfa=0.20

% step (3)
xite1=0.5*alfa*rho*L;
xite2=EI-alfa*EI*L;
xite0_min=max(xite1,xite1*xite2/( xite1-xite2))
xite0=0.10
k_min=xite0
k=20

% step (4)
kd_min=alfa*Ih/(1-alfa)
kd=30

% step (5) Veify other conditions
rho1=3/2*alfa-2*alfa*L^3/EI
rho2=0.5*alfa
rho3=kd-alfa*Ih-kd*alfa
rho4=alfa*kp-kd*alfa-2*alfa*EI*L-2*alfa*L^2
rho5=k-xite0

% get alfa3 and namna
P2=[2*rho1,2*rho2,2*rho3/Ih,2*rho4/kp,2*rho5/m];
alfa3=1.5;
namna0=min(P2)
namna=namna0/alfa3
```

(2) main program: chap5_2.m

```
close all;
clear all;
nx=10;
nt=20000;

tmax=10;L=1;
%Compute mesh spacing and time step
dx=L/(nx-1);
```

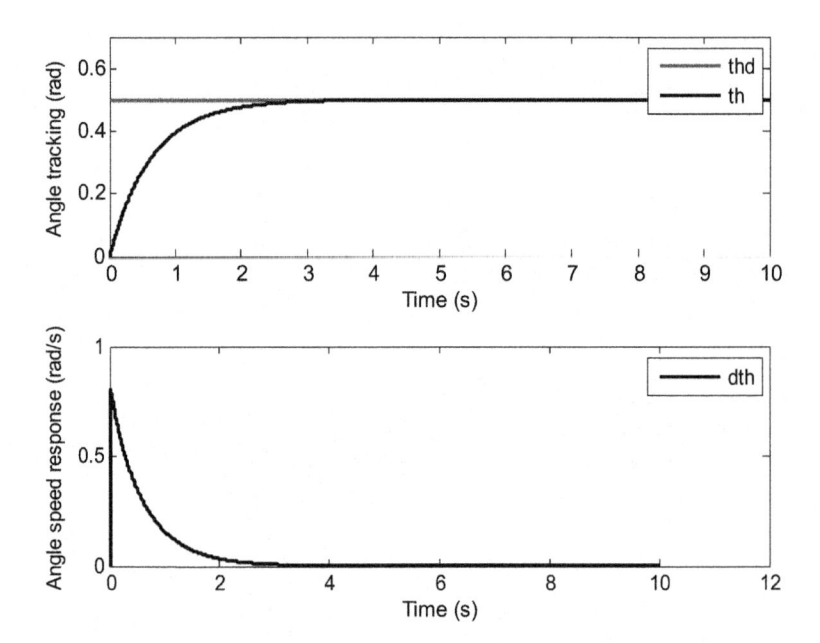

Fig. 5.1 Angle tracking and angle speed tracking

5.4 Simulation Example

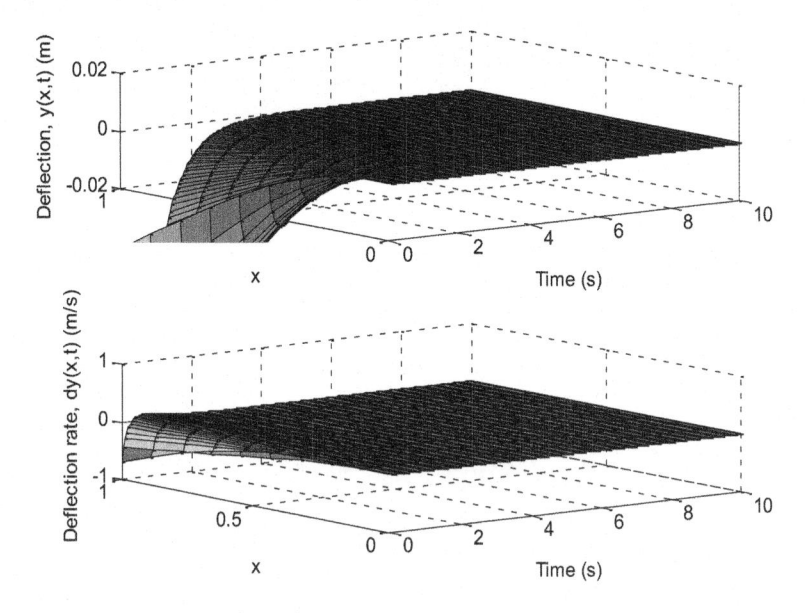

Fig. 5.2 Deformation and deformation rate

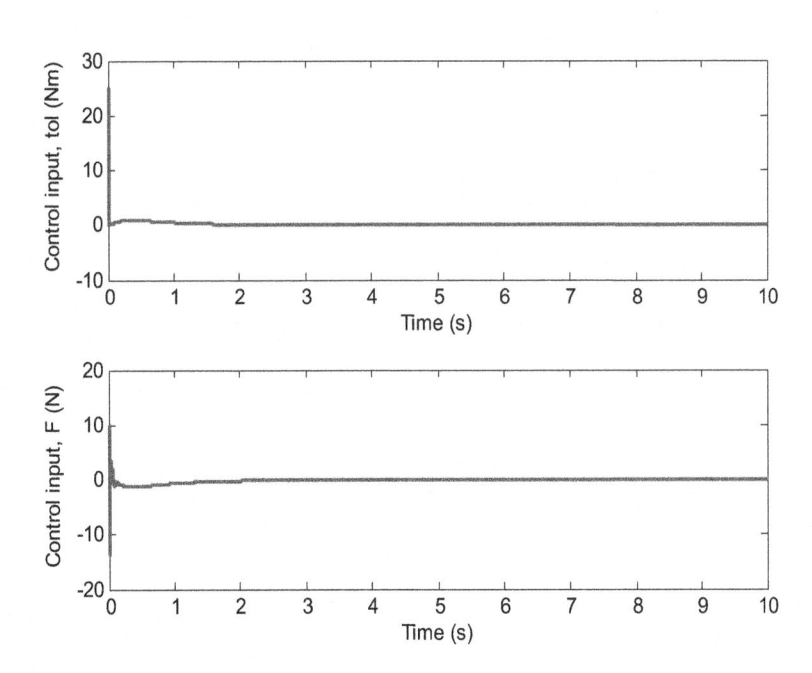

Fig. 5.3 Boundary control input τ and F

```
T=tmax/(nt-1);

%Create arrays to save data for export
t=linspace(0,nt*T,nt);
x=linspace(0,L,nx);

%Parameters
EI=3;rho=0.2;m=0.1;Ih=0.1;
kp=50;kd=30;k=20;

dthd=0;ddthd=0;

dzL_1=0;
zxxxL_1=0;

dzx_1=0;
zxxxx_1=0;
F_1=0;
%Define viriables and Initial condition:
y=zeros(nx,nt);
z=zeros(nx,nt);    %elastic deflectgion
th_2=0;th_1=0;
dth_1=0;

M=2;  %Closed loop
%M=1;  %Open loop
for j=1:nt
    th(j)=0;       %joint angle
    thd(j)=0.5;    %desired angle
    tol(j)=0;
    F(j)=0;
end

for j=3:nt    %Begin
e=th(j-1)-thd(j-1);
de=dth_1-dthd;

tol(j-1)=-kp*e-kd*de;

if M==1
   tol(j-1)=0;
end
if j==10000
   tol(j-1)=10;
```

5.4 Simulation Example

```
end

yxx0=(y(3,j-1)-2*y(2,j-1)+y(1,j-1))/dx^2;
zxx0=yxx0;
th(j)=2*th(j-1)-th(j-2)+T^2/Ih*(tol(j-1)+EI*zxx0);
%%%%%%%%%%%%%%%%%%%%%%%%%%%%%%%%%%%%%%%%%%%%%%%%%%%%%%%%%%%%%%%%
%%%%%%
%z(i,j)
dth(j)=(th(j)-th(j-1))/T;
ddth(j)=(th(j)-2*th(j-1)+th(j-2))/T^2;

%get z(i,j),i=1,2, Boundary conditions  (A2)
y(1,:)=0;     %y(0,t)=0, i=1
y(2,:)=0;     %y(1,t)=0, i=2
z(1,:)=0;     %y(0,t)=0, i=1
z(2,:)=0;     %y(1,t)=0, i=2

%get y(i,j),i=3:nx-2
for i=3:nx-2

yxxxx=(y(i+2,j-1)-4*y(i+1,j-1)+6*y(i,j-1)-4*y(i-1,j-1)+y(i-2,j-1))/dx
^4;
   y(i,j)=T^2*(-i*dx*ddth(j)-(EI*yxxxx)/rho)+2*y(i,j-1)-y(i,j-2);
   zxxxx(i,j-1)=yxxxx;

   dy(i,j-1)=(y(i,j-1)-y(i,j-2))/T;
   dzx(i,j-1)=i*dx*dth(j-1)+dy(j-1);
end

%get z(nx-1,j),i=nx-1
yxxxx(nx-1,j-1)=(-2*y(nx,j-1)+5*y(nx-1,j-1)-4*y(nx-2,j-1)+y(nx-3,j-1)
)/dx^4;
y(nx-1,j)=T^2*(-(nx-1)*dx*ddth(j)-EI*yxxxx(nx-1,j-1)/rho)+2*y(nx-1,j-
1)-y(nx-1,j-2);
zxxxx(nx-1,j-1)=yxxxx(nx-1,j-1);
dy(nx-1,j)=(y(nx-1,j)-y(nx-1,j-1))/T;

%dzx(nx-1,j)=(nx-1)*dx*dth(j)+dyx(nx-1,j);
%get y(nx,j),y=nx
yxxxL(j-1)=(-y(nx,j-1)+2*y(nx-1,j-1)-y(nx-2,j-1))/dx^3;
y(nx,j)=T^2*(-L*ddth(j-1)+(EI*yxxxL(j-1)+F_1)/m)+2*y(nx,j-1)-y(nx,j-2
);

dy(nx,j)=(y(nx,j)-y(nx,j-1))/T;
```

```
zxxxL(j-1)=yxxxL(j-1);

dyL(j-1)=(y(nx,j-1)-y(nx,j-2))/T;
dzL(j-1)=L*dth(j-1)+dyL(j-1);

ua=dzL(j-1)-zxxxL(j-1);
dzxxx_L=(yxxxL(j-1)-yxxxL(j-2))/T;
F(j-1)=-k*ua+m*dzxxx_L;

if M==1
    F(j-1)=0;
end
F_1=F(j-1);
dth_1=dth(j);
dzL_1=dzL(j-1);
zxxxL_1=zxxxL(j-1);
end    %End
%To view the curve, short the points
tshort=linspace(0,tmax,nt/100);
yshort=zeros(nx,nt/100);
dyshort=zeros(nx,nt/100);
for j=1:nt/100
    for i=1:nx
        yshort(i,j)=y(i,j*100);      %Using true y(i,j)
        dyshort(i,j)=dy(i,j*100);    %Using true dy(i,j)
    end
end
figure(1);
subplot(211);
plot(t,thd,'r',t,th,'k','linewidth',2);
xlabel('Time (s)');ylabel('Angle tracking');
legend('thd','th');
axis([0 10 0 0.7]);
subplot(212);
plot(t,dth,'k','linewidth',2);
xlabel('Time (s)');ylabel('Angle speed response)');
legend('dth');

figure(2);
subplot(211);
surf(tshort,x,yshort);
xlabel('Time (s)'); ylabel('x');zlabel('Deflection, y(x,t)');
axis([0 10 0 1 -0.02 0.02]);
subplot(212);
```

5.4 Simulation Example

```
surf(tshort,x,dyshort);
xlabel('Time (s)'); ylabel('x');zlabel('Deflection rate, dy(x,t)');

figure(3);
subplot(211);
plot(t,tol,'r','linewidth',2);
xlabel('Time (s)');ylabel('Control input, tol');
axis([0 10 -10 30]);
subplot(212);
plot(t,F,'r','linewidth',2);
xlabel('Time (s)');ylabel('Control input, F');
axis([0 10 -20 20]);
```

Appendix

Design of k_p, k_d and k

(1) Let $k_p = 50$;
(2) Design α: $0 < \alpha < \dfrac{1}{\max\left(2L, \frac{2\rho L^3}{EI}, \frac{2(I_h + 2\rho L^2)}{k_p}, 2\right)}$;
(3) Design η_0: from $\eta_1 = \frac{1}{2}\alpha\rho L$ and $\eta_2 = EI - \alpha EIL$, we get $\eta_0 > \max\left(\eta_1, \frac{\eta_1 \eta_2}{\eta_2 - \eta_1}\right)$,
(4) From $\sigma_3 = k_d - \alpha I_h - k_d\alpha > 0$ and $\sigma_5 = k - \eta_0 > 0$, let $k_d = 30$, we choose $k = 20$;
(5) Verify
 $\sigma_1 = \frac{3}{2}\alpha - 2\alpha L^2 - \frac{2\alpha L^3}{EI} > 0, \sigma_2 = \frac{1}{2}\alpha > 0, \sigma_4 = \alpha k_p - k_d\alpha - 2\alpha EIL - 2\alpha L^2 > 0, \sigma_5 = k - \eta_0 > 0$
(6) From $2 > \alpha_3 > 1$, Lack of subject $\alpha_3 = 1.5$;
(7) Design λ: from $\min\left(2\sigma_1, 2\sigma_2, \frac{2\sigma_3}{I_h}, \frac{2\sigma_4}{k_p}, \frac{2\sigma_5}{m}\right) \geq \lambda_0 > 0$, we get $\lambda = \frac{\lambda_0}{\alpha_3}$.

References

1. C.D. Rahn, *Mechatronic Control of Distributed Noise and Vibration* (Springer, New York, 2001)
2. G.H. Hardy, J.E. Littlewood, G. Polya, *Inequalities* (Cambridge University Press, Cambridge, 1959)
3. Ioannou, P.A., Sun, J.: Robust Adaptive Control. PTR Prentice-Hall (1996)
4. T.T. Jiang, J.K. Liu, W. He, Boundary Control for a flexible manipulator based on infinite dimensional disturbance observer. J. Sound Vib. **348**(21), 1–14 (2015)

Chapter 6
Boundary Control for Flexible Manipulator with LaSalle Analysis

In Chap. 5, a boundary controller for flexible manipulator with exponential convergence is designed. However, in the controller design, $z_{xxx}(L)$ and $\dot{z}_{xxx}(L)$ are needed. In this chapter, to weaken the vibration of the manipulator, the boundary controller design method based on LaSalle analysis is introduced, where $z_{xxx}(L)$ and $\dot{z}_{xxx}(L)$ are not needed.

6.1 System Description

Considering PDE model given in Chap. 3, the PDE model is given as

$$\rho \ddot{z}(x) = -EI z_{xxxx}(x) \tag{6.1}$$

$$\tau = I_h \ddot{z}_x(0) - EI z_{xx}(0) \tag{6.2}$$

$$F = m\ddot{z}(L) - EI z_{xxx}(L) \tag{6.3}$$

$$z(0) = 0,\ z_x(0) = \theta,\ z_{xx}(L) = 0 \tag{6.4}$$

where $z(x) = x\theta + y(x),\ \ddot{z}(x) = x\ddot{\theta} + \ddot{y}(x),\ \ddot{z}(L) = L\ddot{\theta} + \ddot{y}(L)$.
From $z(x)$ definition, we have

$$z_{xx}(x) = y_{xx}(x), \ddot{z}_x(0) = \ddot{\theta}, z_{xx}(0) = y_{xx}(0), z_{xx}(L) = y_{xx}(L), z_{xxx}(L) = y_{xxx}(L) \tag{6.5}$$

Considering $\theta_d(t)$ as an ideal angle signal, $\theta_d(t)$ is constant value. The control goal is: $\theta(t) \to \theta_d(t),\ \dot{\theta}(t) \to \dot{\theta}_d(t),\ y(x,t) \to 0,\ \dot{y}(x,t) \to 0$.

Define $e = \theta(t) - \theta_d(t)$, then we have

$$\dot{e} = \dot{\theta}(t) - \dot{\theta}_d(t) = \dot{\theta}(t), \quad \ddot{e} = \ddot{\theta}(t) - \ddot{\theta}_d(t) = \ddot{\theta}(t)$$

6.2 Dissipative Analysis of the Closed System

Design controller as

$$\tau = -k_p e - k_d \dot{e} \tag{6.6}$$

$$F = -k\dot{z}(L, t) \tag{6.7}$$

where $k_p > 0$, $k_d > 0$, $k > 0$. F is boundary control input at the end of mechanical arm.

Define

$$q = [q_1 \quad q_2 \quad q_3 \quad q_4 \quad q_5]^T = [e \quad \dot{e} \quad \dot{z}(L,t) \quad y(x,t) \quad \dot{y}(x,t)]^T$$

The closed-loop system can be compactly written as

$$\dot{q} = \mathcal{A}q, \quad q(0) \in \mathcal{H} \tag{6.8}$$

The spaces related to q mentioned above are defined as

$$\mathcal{H} = \mathbb{R}^3 \times H^2 \times L^2 \tag{6.9}$$

where $L^2(\Omega) = \left\{ f \mid \int_\Omega |f(x)|^2 dx < \infty \right\}$, $H^k(\Omega) = \{ f \mid f, f', \ldots, f^{(k)} \in L^2(\Omega) \}$, $\Omega = [0, L]$.

In $\mathcal{H}$, $\mathbb{R}^3$ is defined for q_1, that is, q_1, q_2 and q_3 are all real number. H^2 is defined for q_4, that is, $y(x,t)$, $y_x(x,t)$ and $y_{xx}(x,t)$ are all L^2 limited. L^2 is defined for q_5, that is, $\dot{y}(x,t)$ is L_2 limited.

$\mathcal{A}$ is an infinite dimensional linear operator, define

$$\mathcal{A}q = [\dot{q}_1 \quad \dot{q}_2 \quad \dot{q}_3 \quad \dot{q}_4 \quad \dot{q}_5]^T, \quad \forall q \in \mathcal{D}(\mathcal{A}) \tag{6.10}$$

From (6.2), we have $I_h \ddot{\theta}(t) = \tau + EI y_{xx}(0, t)$. From (6.6), we have $\ddot{e} = \frac{1}{I_h}(-k_p q_1 - k_d q_2 + EI y_{xx}(0, t))$, then we get

6.2 Dissipative Analysis of the Closed System

$$\ddot{e} = \frac{1}{I_h}\left(-k_p q_1 - k_d q_2 + EIq_{4,xx}(0)\right)$$

From $m\left(L\ddot{\theta}(t) + \ddot{y}(L,t)\right) = EIy_{xxx}(L,t) + F$, we have $\dot{q}_3 = \ddot{z}(L,t) = \frac{1}{m}(EIy_{xxx}(L,t) + F) = \frac{1}{m}\left(EIq_{4,xxx}(L) - kq_3\right)$, then we get

$$\ddot{z}(L,t) = \frac{1}{m}\left(EIq_{4,xxx}(L) - kq_3\right)$$

From $\rho\left(x\ddot{\theta}(t) + \ddot{y}(x,t)\right) = -EIy_{xxxx}(x,t)$, we have $\ddot{y}(x,t) = -\frac{1}{\rho}EIy_{xxxx}(x,t) - x\ddot{\theta}(t)$, then

$$\ddot{y}(x,t) = -\frac{1}{\rho}EIq_{4,xxxx} - x\vartheta_1 = -\frac{1}{\rho}EIq_{4,xxxx} - \frac{x}{I_h}\left(-k_p q_1 - k_d q_2 + EIq_{4,xx}(0)\right)$$

Then we get

$$\mathcal{A}q = \begin{bmatrix} q_2 \\ \frac{1}{I_h}\left(-k_p q_1 - k_d q_2 + EIq_{4,xx}(0)\right) \\ \frac{1}{m}\left(EIq_{4,xxx}(L) - kq_3\right) \\ q_5 \\ -\frac{1}{\rho}EIq_{4,xxxx} - \frac{x}{I_h}\left(-k_p q_1 - k_d q_2 + EIq_{4,xx}(0)\right) \end{bmatrix} \quad (6.11)$$

In $q = [e \quad \dot{e} \quad \dot{z}(L,t) \quad y(x,t) \quad \dot{y}(x,t)]^T$ and $\mathcal{D}(\mathcal{A})$, we design Lyapunov function as

$$V = E_1 + E_2 \quad (6.12)$$

where $E_1 = \frac{1}{2}\int_0^L \rho\dot{z}^2(x,t)dx + \frac{1}{2}EI\int_0^L y_{xx}^2(x,t)dx$, $E_2 = \frac{1}{2}I_h\dot{e}^2 + \frac{1}{2}k_p e^2 + \frac{1}{2}m\dot{z}^2(L,t)$, $k_p > 0$.

In above Lyapunov function, the manipulator kinetic energy $\frac{1}{2}\int_0^L \rho\dot{z}^2(x,t)dx$, the manipulator potential energy $\frac{1}{2}EI\int_0^L y_{xx}^2(x,t)dx$, and the load kinetic energy are all considered.
Then

$$\dot{V} = \dot{E}_1 + \dot{E}_2$$

where

$$\begin{aligned}
\dot{E}_1 &= \int_0^L \rho \dot{z}(x,t)\ddot{z}(x,t)dx + EI\int_0^L y_{xx}(x,t)\dot{y}_{xx}(x,t)dx \\
&= \int_0^L -EIy_{xxxx}(x,t)\dot{z}(x,t)dx + EIy_{xx}(x,t)\dot{y}_x(x,t)|_0^L - \int_0^L EIy_{xxx}(x,t)\dot{y}_x(x,t)dx \\
&= \int_0^L -EIy_{xxxx}(x,t)\dot{z}(x,t)dx + EIy_{xx}(L,t)\dot{y}_x(L,t) - EIy_{xx}(0,t)\dot{y}_x(0,t) \\
&\quad - (EIy_{xxx}(x,t)\dot{y}(x,t)|_0^L - \int_0^L EIy_{xxxx}(x,t)\dot{y}(x,t)dx) \\
&= \int_0^L -EIy_{xxxx}(x,t)(x\dot{\theta}(t)+\dot{y}(x,t))dx + \int_0^L EIy_{xxxx}(x,t)\dot{y}(x,t)dx - EIy_{xxx}(L,t)\dot{y}(L,t) + EIy_{xxx}(0,t)\dot{y}(0,t) \\
&= \int_0^L -EIy_{xxxx}(x,t)x\dot{\theta}(t)dx - EIy_{xxx}(L,t)\dot{y}(L,t) \\
&= -EIy_{xxx}(x,t)x\dot{\theta}(t)|_0^L - \int_0^L -EIy_{xxx}(x,t)\dot{\theta}(t)dx - EIy_{xxx}(L,t)\dot{y}(L,t) \\
&= -EIy_{xxx}(L,t)L\dot{\theta}(t) + \int_0^L EIy_{xxx}(x,t)\dot{\theta}(t)dx - EIy_{xxx}(L,t)\dot{y}(L,t) \\
&= -EIy_{xxx}(L,t)L\dot{\theta}(t) + EIy_{xx}(x,t)\dot{\theta}(t)|_0^L - EIy_{xxx}(L,t)\dot{y}(L,t) \\
&= -EIy_{xxx}(L,t)L\dot{\theta}(t) + EIy_{xx}(L,t)\dot{\theta}(t) - EIy_{xx}(0,t)\dot{\theta}(t) - EIy_{xxx}(L,t)\dot{y}(L,t) \\
&= -EIy_{xxx}(L,t)L\dot{\theta}(t) - EIy_{xx}(0,t)\dot{\theta}(t) - EIy_{xxx}(L,t)\dot{y}(L,t) \\
&= -EIy_{xxx}(L,t)\dot{z}(L,t) - EIy_{xx}(0,t)\dot{\theta}(t)
\end{aligned}$$

$$\dot{E}_2 = I_h\ddot{e}\dot{e} + k_p e\dot{e} + m\dot{z}(L,t)\ddot{z}(L,t) = \dot{e}(I_h\ddot{e}+k_p e) + \dot{z}(L,t)m\ddot{z}(L,t)$$

Therefore

$$\begin{aligned}
\dot{V} &= \dot{E}_1 + \dot{E}_2 \\
&= -EIy_{xxx}(L,t)\dot{z}(L,t) - EIy_{xx}(0,t)\dot{\theta}(t) + \dot{e}(I_h\ddot{e}+k_p e) + \dot{z}(L,t)m\ddot{z}(L,t) \\
&= \dot{e}(I_h\ddot{e}+k_p e - EIy_{xx}(0,t)) + \dot{z}(L,t)(-EIy_{xxx}(L,t) + m\ddot{z}(L,t)) \\
&= \dot{e}\left(I_h \cdot \frac{1}{I_h}(\tau + EIy_{xx}(0,t)) + k_p e - EIy_{xx}(0,t)\right) + \dot{z}(L,t)\left(-EIy_{xxx}(L,t) + m\frac{1}{m}(EIy_{xxx}(L,t)+F)\right) \\
&= \dot{e}(\tau + k_p e) + \dot{z}(L,t)F
\end{aligned}$$

Submitting controller into above, then

$$\dot{V} = -k_d\dot{e}^2 - k\dot{z}^2(L,t) \leq 0$$

Hence, the operator $\mathcal{A}$ is dissipative.

6.3 Unique Analysis of Solutions

To prove the unique solutions of q and $\mathcal{A}^{-1}$ is a compact operator, define $g = [g_1 \ g_2 \ g_3 \ g_4 \ g_5]^T \in \mathcal{H}$, let

$$\mathcal{A}q = g \qquad (6.13)$$

From,

$$\mathcal{A}q = \begin{bmatrix} q_2 \\ \frac{1}{I_h}\left(-k_p q_1 - k_d q_2 + EI q_{4,xx}(0)\right) \\ \frac{1}{m}\left(EI q_{4,xxx}(L) - k q_3\right) \\ q_5 \\ -\frac{1}{\rho} EI q_{4,xxxx} - \frac{x}{I_h}\left(-k_p q_1 - k_d q_2 + EI q_{4,xx}(0)\right) \end{bmatrix}$$

we have

$$g_1 = q_2$$
$$g_2 = \frac{1}{I_h}\left(-k_p q_1 - k_d q_2 + EI q_{4,xx}(0)\right)$$
$$g_3 = \frac{1}{m}\left(EI q_{4,xxx}(L) - k q_3\right)$$
$$g_4 = q_5$$
$$g_5 = -\frac{1}{\rho} EI q_{4,xxxx} - \frac{x}{I_h}\left(-k_p q_1 - k_d q_2 + EI q_{4,xx}(0)\right)$$

The solution of $g_5 = -\frac{1}{\rho} EI q_{4,xxxx} - \frac{x}{I_h}\left(-k_p q_1 - k_d q_2 + EI q_{4,xx}(0)\right)$ is

$$q_4 = -\frac{\rho}{EI}\int_0^x\int_0^{\xi_1}\int_0^{\xi_2}\int_0^{\xi_3} g_5(\xi_4) d\xi_4 d\xi_3 d\xi_2 d\xi_1 - \frac{x^5}{5!}\frac{\rho}{EI} g_2 + \sum_{j=0}^{3} \sigma_j x^j$$

where $\sigma_0, \ldots, \sigma_3$ are uniquely determined by boundary conditions (6.3) and (6.4). Then we can get the unique solution as

$$q_1 = \frac{1}{k_p}\left(-I_h g_2 - k_d g_1 + EI q_{4,xx}(0)\right)$$
$$q_2 = g_1$$
$$q_3 = \frac{1}{k}\left(EI q_{4,xxx}(L) - m g_3\right) \qquad (6.14)$$
$$q_4 = -\frac{\rho}{EI}\int_0^x\int_0^{\xi_1}\int_0^{\xi_2}\int_0^{\xi_3} g_5(\xi_4) d\xi_4 d\xi_3 d\xi_2 d\xi_1 - \frac{x^5}{5!}\frac{\rho}{EI} g_2 + \sum_{j=0}^{3} \sigma_j x^j$$
$$q_5 = g_4$$

Hence, equation $\mathcal{A}q = g$ has a unique solution $q \in \mathcal{D}(\mathcal{A})$, implying that $\mathcal{A}^{-1}$ exists and maps $\mathcal{H}$ into $\mathcal{H} = \mathbb{R}^3 \times H^2 \times L^2$. Moreover, since $\mathcal{A}^{-1}$ maps every bounded set of $\mathcal{H}$ into bounded set of $\mathcal{H} = \mathbb{R}^3 \times H^2 \times L^2$, the embedding of the later space onto $\mathcal{H}$ is compact. It follows that $\mathcal{A}^{-1}$ is a compact operator.

The spectrum of $\mathcal{A}$ consists entirely of isolated eigen-values. It also proves that for any $\lambda > 0$ in the resolvent set of $\mathcal{A}$, the operator $(\lambda I - \mathcal{A})^{-1}$ is a compact operator. Based on the Lumer-Phillips theorem, operator $\mathcal{A}$ generates a C_0-semigroup of contractions $T(t)$ on $\mathcal{H}$ [1].

6.4 Convergence Analysis

Let $\dot{V} \equiv 0$, we have $\dot{e} \equiv \dot{z}(L, t) \equiv 0$, $\ddot{e} \equiv \ddot{z}(L, t) \equiv 0$. Considering θ_d is constant value, from $\dot{e} = \dot{\theta}(t)$, $\ddot{e} = \ddot{\theta}(t)$, we have $\dot{\theta}(t) \equiv 0$, $\ddot{\theta}(t) \equiv 0$. From $\rho\left(x\ddot{\theta}(t) + \ddot{y}(x,t)\right) = \rho\ddot{z}(x,t) = -EIy_{xxxx}(x,t)$, we have

$$\rho\ddot{y}(x,t) = -EIy_{xxxx}(x,t),$$

$$\rho\ddot{z}(L,t) = -EIy_{xxxx}(L,t) = 0$$

then we get $y_{xxxx}(L,t) = 0$.

Using the technique of separation of variables [2]. We can write $y(x,t)$ as follows:

$$y(x,t) = X(x) \cdot T(t) \tag{6.15}$$

where $X(x)$ and $T(t)$ are unknown functions.

From $\rho\ddot{y}(x,t) = -EIy_{xxxx}(x,t)$, we have

$$y_{xxxx}(x,t) = -\frac{\rho}{EI}\ddot{y}(x,t)$$

From (6.15), we have $y_{xxxx}(x,t) = X^{(4)}(x) \cdot T(t)$, $\ddot{y}(x,t) = X(x) \cdot T^{(2)}(t)$, then above equation becomes

$$\frac{X^{(4)}(x)}{X(x)} = -\frac{\rho}{EI}\frac{T^{(2)}(t)}{T(t)} = \mu$$

i.e.

$$X^{(4)}(x) - \mu X(x) = 0$$

6.4 Convergence Analysis

Let $\mu = \beta^4$, then we can get the solution as

$$X(x) = c_1 \cosh \beta x + c_2 \sinh \beta x + c_3 \cos \beta x + c_4 \sin \beta x \quad (6.16)$$

where $c_i \in R$, $i = 1, 2, 3, 4$ are unknown real number.

Considering $y(0,t) = 0$, $y_x(0,t) = 0$, $y_{xx}(L,t) = 0$ and $y_{xxxx}(L,t) = 0$, combining with (6.15), we have $X(0) = X'(0) = X''(L) = X^{(4)}(L) = 0$. Then from (6.16), we have

$$\begin{cases} c_1 + c_3 = 0 \\ c_2 + c_4 = 0 \\ c_1 \cosh \beta L + c_2 \sinh \beta L - c_3 \cos \beta L - c_4 \sin \beta L = 0 \\ c_1 \cosh \beta L + c_2 \sinh \beta L + c_3 \cos \beta L + c_4 \sin \beta L = 0 \end{cases} \quad (6.17)$$

Then we have

$$\begin{cases} c_1 \cosh \beta L + c_2 \sinh \beta L = 0 \\ c_3 \cos \beta L + c_4 \sin \beta L = 0 \end{cases}$$

i.e.

$$\begin{cases} c_3 \cosh \beta L + c_4 \sinh \beta L = 0 \\ c_3 \cos \beta L + c_4 \sin \beta L = 0 \end{cases}$$

therefore

$$c_4(\sinh \beta L \cdot \cos \beta L - \sin \beta L \cdot \cosh \beta L) = 0$$

We can conclude that $X^{(4)}(x) - \mu X(x) = 0$ have unique solutions, $c_i = 0$, $i = 1, 2, 3, 4$, thus, $X(x) = 0$, $y(x,t) = 0$ and $X_{xx}(0) = -c_1 + c_3 = 0$.
From $y(x,t) = X(x) \cdot T(t)$, we have $y_{xx}(0) = X_{xx}(0)T(t) = 0$.
Consider

$$I_h \ddot{\theta}(t) = \tau + EI y_{xx}(0,t) = -k_p e - k_d \dot{e} + EI y_{xx}(0)$$

Let $\dot{V} \equiv 0$, thus $\dot{e} \equiv 0$, $\ddot{\theta} \equiv 0$, combining with $y_{xx}(0) = 0$, then we have $e = 0$. Therefore, according to the extended LaSalle's invariance principle [3], the PDE boundary control (6.6) and (6.7) can guarantee the asymptotic stability of the closed-loop system. If then, $t \to \infty$, $e \to 0$, $\dot{e} \to 0$, $y(x,t) \to 0$.

6.5 Simulation Example

Considering the PDE model as Eqs. (6.1)–(6.5), the sampling time is set as $\Delta t = 5 \times 10^{-4}$, and set $\Delta x = 0.01$, the physical parameters are chosen as: $EI = 3.0$, $L = 1.0$, $\rho = 0.20$, $m = 0.10$, $I_h = 0.10$.

Define ideal angle as $\theta_d = 0.5$, use controller (6.6) and (6.7), let $\theta_d = 0.50$, $k_p = 50$, $k_d = 30$, $k = 20$. Two axes are divided into sections according to $nx = 10$, $nt = 20000$. The simulation results are shown from Figs. 6.1, 6.2 and 6.3.

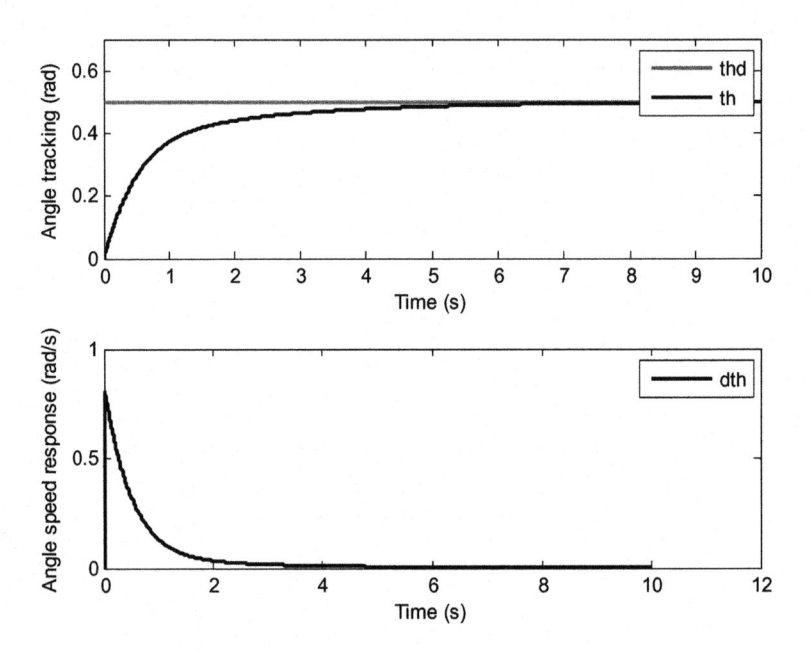

Fig. 6.1 Angle tracking and angle speed tracking

6.5 Simulation Example

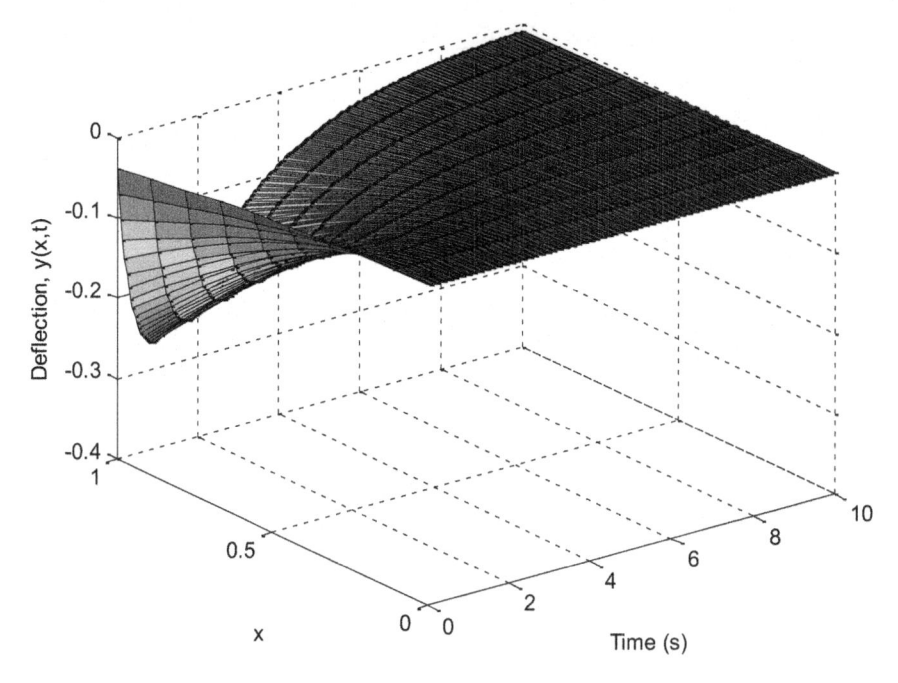

Fig. 6.2 Deformation $y(x,t)$

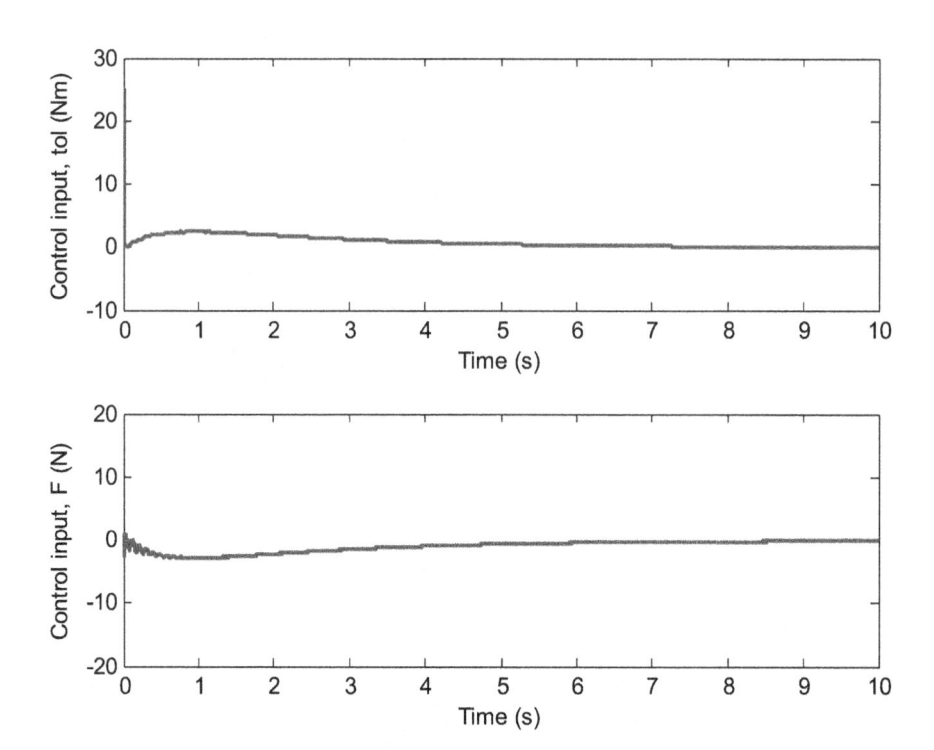

Fig. 6.3 Boundary control input, τ and F

Simulation program: chap6_1.m
close all;
clear all;
nx=10;
nt=20000;

tmax=10;L=1;
dx=L/(nx-1);
T=tmax/(nt-1);

t=linspace(0,nt*T,nt);
x=linspace(0,L,nx);

EI=3;rho=0.2;m=0.1;Ih=0.1;
kp=50;kd=30;k=20;

dthd=0;ddthd=0;

dzL_1=0;
zxxxL_1=0;

dzx_1=0;
zxxxx_1=0;
F_1=0;
y=zeros(nx,nt);
z=zeros(nx,nt);
th_2=0;th_1=0;
dth_1=0;

for j=1:nt
    th(j)=0;
    thd(j)=0.5;
    tol(j)=0;
    F(j)=0;
end

6.5 Simulation Example

```
for j=3:nt
e=th(j-1)-thd(j-1);
de=dth_1-dthd;

tol(j-1)=-kp*e-kd*de;

yxx0=(y(3,j-1)-2*y(2,j-1)+y(1,j-1))/dx^2;
zxx0=yxx0;
th(j)=2*th(j-1)-th(j-2)+T^2/Ih*(tol(j-1)+EI*zxx0);
dth(j)=(th(j)-th(j-1))/T;
ddth(j)=(th(j)-2*th(j-1)+th(j-2))/T^2;

%Boundary conditions
y(1,:)=0;      %y(0,t)=0, i=1
y(2,:)=0;      %y(1,t)=0, i=2
z(1,:)=0;      %y(0,t)=0, i=1
z(2,:)=0;      %y(1,t)=0, i=2

%get y(i,j),i=3:nx-2
for i=3:nx-2
    yxxxx=(y(i+2,j-1)-4*y(i+1,j-1)+6*y(i,j-1)-4*y(i-1,j-1)+y(i-2,j-1))/dx^4;
    y(i,j)=T^2*(-i*dx*ddth(j)-(EI*yxxxx)/rho)+2*y(i,j-1)-y(i,j-2);      %见式(A3)
    zxxxx(i,j-1)=yxxxx;

    dy(i,j-1)=(y(i,j-1)-y(i,j-2))/T;
    dzx(i,j-1)=i*dx*dth(j-1)+dy(j-1);
end

%get z(nx-1,j),i=nx-1
yxxxx(nx-1,j-1)=(-2*y(nx,j-1)+5*y(nx-1,j-1)-4*y(nx-2,j-1)+y(nx-3,j-1))/dx^4;
y(nx-1,j)=T^2*(-(nx-1)*dx*ddth(j)-EI*yxxxx(nx-1,j-1)/rho)+2*y(nx-1,j-1)-y(nx-1,j-2);
zxxxx(nx-1,j-1)=yxxxx(nx-1,j-1);
```

```
dy(nx-1,j)=(y(nx-1,j)-y(nx-1,j-1))/T;

%get y(nx,j),y=nx
yxxxL(j-1)=(-y(nx,j-1)+2*y(nx-1,j-1)-y(nx-2,j-1))/dx^3;
y(nx,j)=T^2*(-L*ddth(j-1)+(EI*yxxxL(j-1)+F_1)/m)+2*y(nx,j-1)-y(nx,j-2);
dy(nx,j)=(y(nx,j)-y(nx,j-1))/T;
zxxxL(j-1)=yxxxL(j-1);

dyL(j-1)=(y(nx,j-1)-y(nx,j-2))/T;
dzL(j-1)=L*dth(j-1)+dyL(j-1);

dzxxx_L=(yxxxL(j-1)-yxxxL(j-2))/T;

F(j-1)=-k*dzL(j-1);


F_1=F(j-1);
dth_1=dth(j);
dzL_1=dzL(j-1);
zxxxL_1=zxxxL(j-1);
end
tshort=linspace(0,tmax,nt/100);
yshort=zeros(nx,nt/100);
for j=1:nt/100
for i=1:nx
            yshort(i,j)=y(i,j*100);
end
end
figure(1);
subplot(211);
plot(t,thd,'r',t,th,'k','linewidth',2);
xlabel('Time (s)');ylabel('Angle tracking (rad)');
legend('thd','th');
axis([0 10 0 0.7]);
subplot(212);
plot(t,dth,'k','linewidth',2);
xlabel('Time (s)');ylabel('Angle speed response (rad/s)');
legend('dth');
```

```
figure(2);
surf(tshort,x,yshort);
xlabel('Time (s)'); ylabel('x');zlabel('Deflection, y(x,t) (m)');

figure(3);
subplot(211);
plot(t,tol,'r','linewidth',2);
xlabel('Time (s)');ylabel('Control input, tol (Nm)');
axis([0 10 -10 30]);
subplot(212);
plot(t,F,'r','linewidth',2);
xlabel('Time (s)');ylabel('Control input, F (N)');
axis([0 10 -20 20]);
```

References

1. T.T. Jiang, J.K. Liu, W. He, Boundary control for a flexible manipulator based on infinite dimensional disturbance observer. J. Sound Vib. **348**(21), 1–14 (2015)
2. W.H. Ray, *Advanced Process Control* (McGraw-Hill, New York, 1981)
3. C.D. Rahn, *Mechatronic Control of Distributed Noise and Vibration-A Lyapunov Approach* (Springer, Berlin Heidelberg, 2001)

Chapter 7
Boundary Control for Flexible Manipulator with State Constraints

7.1 Introduction

Although the control laws design for flexible manipulator systems has made great progress, studies of flexible manipulators or PDE model systems with output constraints are limited. Constraints are ubiquitous in physical systems and violation of the constraints during operation may lead to performance degradation, hazards or system damage.

It is noted that previous studies about control schemes for flexible manipulators don't consider the problems of position and velocity constraints at the same time. Unlike the previous papers.

In [1], the authors use Barrier Lyapunov Function to design the boundary controller. Under the designed control law, the output information of the flexible manipulator can be limited in a bound, and the closed-loop system is stable.

7.2 System Statement

In this chapter, we consider a flexible manipulator that only moves in the planar plane. The system is shown in Fig. 7.1.

In Fig. 7.1, XOY and xOy donate global inertial coordinate system and the body-fixed coordinate system attached to the manipulator, respectively. The parameter descriptions are given as follows: EI is the bending stiffness, ρ donates the mass per unit length of the flexible manipulator, I_h represents the hub inertia, m is the mass of the payload, L is the length of the manipulator, $\theta(t)$ donates the joint angle, $y(x,t)$ is the vibratory deflection of the link at x, $\tau(t)$ represents the torque input generated by joint motor, $F(t)$ is the force input generated by the actuator, at the tip of the link.

Fig. 7.1 Configuration of the flexible manipulator with state constraints

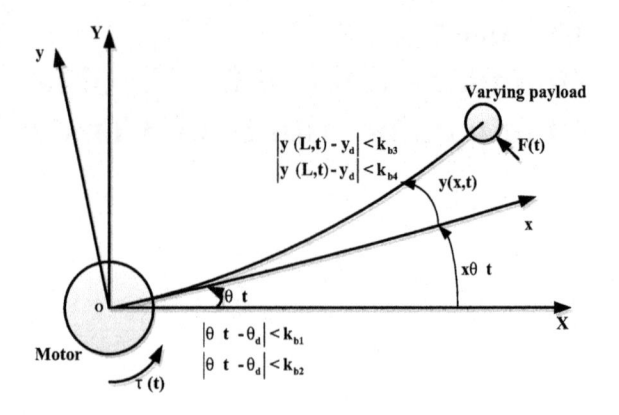

Considering PDE model given in Chap. 3, neglecting disturbance, the PDE model is given as

$$\rho \ddot{z}(x,t) = -EI y_{xxxx}(x,t) \tag{7.1}$$

$$\tau(t) = I_h \ddot{\theta}(t) - EI y_{xx}(0,t) \tag{7.2}$$

$$F(t) = m\ddot{z}(L,t) - EI y_{xxx}(L,t) \tag{7.3}$$

$$y(0,t) = y_x(0,t) = y_{xx}(L,t) = 0 \tag{7.4}$$

Define θ_d as the desired angular position, θ_d is a constant, $e_1 = \theta(t) - \theta_d$, define $y_d(L,t)$ as desired vibration, $y_d(L,t) = 0$, $e_2 = y(L,t) - y_d(L,t) = y(L,t)$.
Then we have $\dot{e}_1 = \dot{\theta}(t) - \dot{\theta}_d = \dot{\theta}(t)$, $\ddot{e}_1 = \ddot{\theta}(t) - \ddot{\theta}_d = \ddot{\theta}(t)$, $\dot{e}_2 = \dot{y}(L,t)$, $\ddot{e}_2 = \ddot{y}(L,t)$.
Define $z(x) = x\theta + y(x)$, then $\ddot{z}(x) = x\ddot{\theta} + \ddot{y}(x)$, $\ddot{z}(L) = L\ddot{\theta} + \ddot{y}(L)$.

7.3 Controller Design and Analysis

Assumption 7.1 There are positive constants $k_{b1}, k_{b2}, k_{b3}, k_{b4}$ satisfying that

$$|e_1(0)| = |\theta(0) - \theta_d| < k_{b1} \tag{7.5}$$

$$|\dot{e}_1(0)| = |\dot{\theta}(0) - \dot{\theta}_d| < k_{b2} \tag{7.6}$$

$$|e_2(0)| = |y(L,0) - y_d(L,0)| < k_{b3} \tag{7.7}$$

7.3 Controller Design and Analysis

$$|\dot{e}_2(0)| = |\dot{y}(L,0) - \dot{y}_d(L,0)| < k_{b4} \tag{7.8}$$

Lemma 7.1 [2] *For any positive constant k_b, let $Z_1 := \{z_1 \in \mathbb{R} : -k_b < z_1 < k_b\} \subset \mathbb{R}$ and $N := \mathbb{R}^l \times Z_1 \subset \mathbb{R}^{l+1}$ be open sets. Consider the system*

$$\dot{\eta} = h(t, \eta)$$

where $\eta := [w, z_1]^T \in N$, and $h : \mathbb{R}_+ \times N \to \mathbb{R}^{l+1}$ is piecewise continuous in t and locally Lipschitz in z, uniformly in t, on $\mathbb{R}_+ \times N$. Suppose that exist functions $U : \mathbb{R}^l \to \mathbb{R}_+$ and $V_1 : Z_1 \to \mathbb{R}_+$, continuously differentiable and positive definite in their respective domains, such that

$$V_1(z_1) \to \infty \quad \text{as} \quad z_1 \to -k_b \quad \text{or} \quad z_1 \to k_b \tag{7.9}$$

$$\gamma_1(\|w\|) \leq U(w) \leq \gamma_2(\|w\|)$$

where γ_1 and γ_2 are class K_∞ functions. Let $V(\eta) := V_1(z_1) + U(w)$, and $z_1(0)$ belong to the set $z_1 \in (-k_b, k_b)$. If the inequality holds:

$$\dot{V} = \frac{\partial V}{\partial \eta} h \leq 0 \tag{7.10}$$

then $z_1(t)$ remains in the open set $z_1 \in (-k_b, k_b) \ \forall t \in [0, \infty)$.

For example, consider a kind of symmetrical Barrier Lyapunov as

$$V = \frac{1}{2} \log \frac{k_b^2}{k_b^2 - z_1^2}$$

where $\log(\bullet)$ is natural logarithm.

It can be seen that the above Lyapunov function meets $V(0) = 0$, $V(x) > 0 (x \neq 0)$.

For example, if we choose $z_1(0) = 0.5$, consider $|z_1(0)| < k_b$, we can set $k_b = 0.51$. 'The Barrier Lyapunov Function' $V = \frac{1}{2} \log \frac{k_b^2}{k_b^2 - z_1^2}$ is illustrated in Fig. 7.2.

The control objective is to design boundary controller based on a Barrier Lyapunov Function to stabilize the flexible manipulator at the desired angle, suppress vibration and achieve output regulation with state constraints.

It's supposed that the following parameters can be obtained: $y_{xx}(0,t)$, $y_{xxx}(L,t)$. The control inputs are designed as

$$\tau(t) = -EIy_{xx}(0,t) + \left(\frac{k_{b2}^2 - \dot{e}_1^2}{k_{b2}^2 - \dot{e}_1^2 + 1}\right)\left(-\frac{e_1}{k_{b1}^2 - e_1^2} + EILy_{xxx}(L,t) + EIy_{xx}(0,t) - k_1 e_1 - k_3 \dot{e}_1\right) \tag{7.11}$$

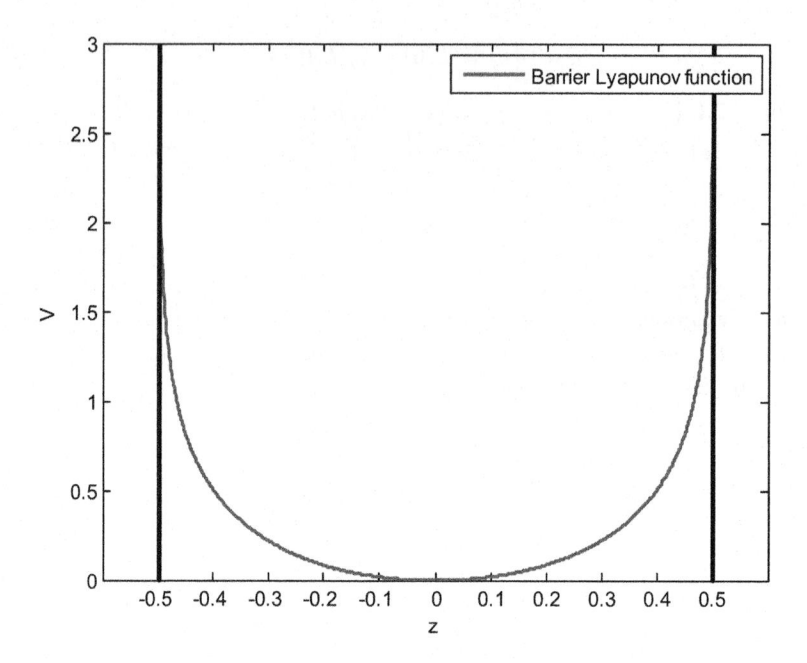

Fig. 7.2 A symmetrical Barrier Lyapunov Function

$$F(t) = -EIy_{xxx}(L,t) + \alpha L(\tau(t) + EIy_{xx}(0,t)) + \left(\frac{k_{b4}^2 - \dot{e}_2^2}{k_{b4}^2 - \dot{e}_2^2 + 1}\right) \left(-\frac{e_2}{k_{b3}^2 - e_2^2} + EIy_{xxx}(L,t) - k_2 e_2 - k_4 \dot{e}_2\right) \quad (7.12)$$

where k_1, k_2, k_3 and k_4 are any positive constants, $\alpha = \frac{m}{I_h}$.

Theorem 7.1 *Considering the system* (7.1)–(7.4) *and boundary control laws* (7.11)–(7.12), *the following properties hold.*

(1) *The closed-loop system is asymptotically stable in the following sense:* $\theta(t) \to \theta_d$, $\dot{\theta}(t) \to \dot{\theta}_d$, $y(x,t) \to 0$, $y(L,t) \to 0$, $\dot{y}(L,t) \to 0$, *as* $x \in [0,L]$, $t \to \infty$.

(2) *Provided that the initial conditions satisfy the Assumption* 7.1, *based on a Barrier Lyapunov Function, we can obtain that* $|e_1| < k_{b1}$, $|\dot{e}_1| < k_{b2}$, $|e_2| < k_{b3}$, $|\dot{e}_2| < k_{b4}$. *Further, we can conclude that* $-k_{b1} + \theta_d < \theta(t) < k_{b1} + \theta_d$, $-k_{b2} < \dot{\theta}(t) < k_{b2}$, $-k_{b3} + y_d(L,t) < y(L,t) < k_{b3} + y_d(L,t)$, $-k_{b4} < \dot{y}(L,t) < k_{b4}$.

Proof Considering $z(x) = x\theta + y(x)$, from (7.2), (7.3), we get

7.3 Controller Design and Analysis

$$m\ddot{y}(L,t) = F(t) + EIy_{xxx}(L,t) - mL\ddot{\theta}(t)$$
$$= F(t) + EIy_{xxx}(L,t) - mL\frac{1}{I_h}(\tau(t) + EIy_{xx}(0,t)) \quad (7.13)$$
$$= F(t) + EIy_{xxx}(L,t) - \alpha L(\tau(t) + EIy_{xx}(0,t))$$

Choose the following Lyapunov function candidate

$$V(t) = V_1(t) + V_2(t) + V_3(t) + V_4(t) + V_5(t) \quad (7.14)$$

in which

$$V_1(t) = \frac{1}{2}\ln\frac{k_{b1}^2}{k_{b1}^2 - e_1^2} + \frac{I_h}{2}\ln\frac{k_{b2}^2}{k_{b2}^2 - \dot{e}_1^2}$$

$$V_2(t) = \frac{1}{2}\ln\frac{k_{b3}^2}{k_{b3}^2 - e_2^2} + \frac{m}{2}\ln\frac{k_{b4}^2}{k_{b4}^2 - \dot{e}_2^2}$$

$$V_3(t) = \frac{1}{2}\int_0^L \rho\dot{z}^2(x,t)dx + \frac{1}{2}EI\int_0^L y_{xx}^2(x,t)dx$$

$$V_4(t) = \frac{1}{2}k_1 e_1^2 + \frac{1}{2}k_2 e_2^2 + \frac{1}{2}I_h\dot{e}_1^2 + \frac{1}{2}m\dot{e}_2^2$$

where $\gamma > 0$.

The derivative of (7.14) is given by

$$\dot{V}(t) = \dot{V}_1(t) + \dot{V}_2(t) + \dot{V}_3(t) + \dot{V}_4(t) + \dot{V}_5(t)$$

where

$\dot{V}_1(t) = \frac{e_1\dot{e}_1}{k_{b1}^2 - e_1^2} + \frac{I_h\dot{e}_1\ddot{e}_1}{k_{b2}^2 - \dot{e}_1^2}$,

$\dot{V}_2(t) = \frac{e_2\dot{e}_2}{k_{b3}^2 - e_2^2} + \frac{m\dot{e}_2\ddot{e}_2}{k_{b4}^2 - \dot{e}_2^2}$,

$\dot{V}_3(t) = \int_0^L \rho\dot{z}(x,t)\ddot{z}(x,t)dx + EI\int_0^L y_{xx}(x,t)\dot{y}_{xx}(x,t)dx$

$\dot{V}_4(t) = k_1 e_1\dot{e}_1 + k_2 e_2\dot{e}_2 + I_h\dot{e}_1\ddot{e}_1 + m\dot{e}_2\ddot{e}_2$.

Applying the system model equations, the boundary conditions, we obtain

$$\dot{V}_3(t) = \int_0^L -EIy_{xxxx}(x,t)\dot{z}(x,t)dx + EI\int_0^L y_{xx}(x,t)\dot{y}_{xx}(x,t)dx$$
$$= -EIy_{xxx}(L,t)L\dot{\theta}(t) - EIy_{xx}(0,t)\dot{\theta}(t) - EIy_{xxx}(L,t)\dot{y}(L,t)$$
$$= \dot{e}_1(-EILy_{xxx}(L,t) - EIy_{xx}(0,t)) - EIy_{xxx}(L,t)\dot{e}_2$$

Then, we have

$$\dot{V}(t) = \dot{e}_1 \left(\frac{e_1}{k_{b1}^2 - e_1^2} - EILy_{xxx}(L,t) - EIy_{xx}(0,t) + k_1 e_1 + \frac{I_h \ddot{e}_1}{k_{b2}^2 - \dot{e}_1^2} + I_h \ddot{e}_1 \right)$$

$$+ \dot{e}_2 \left(\frac{e_2}{k_{b3}^2 - e_2^2} - (EIy_{xxx}(L,t) - k_2 e_2) + \frac{m\ddot{e}_2}{k_{b4}^2 - \dot{e}_2^2} + m\ddot{e}_2 \right)$$

$$= \dot{e}_1 \left(\frac{e_1}{k_{b1}^2 - e_1^2} - EILy_{xxx}(L,t) - EIy_{xx}(0,t) + k_1 e_1 + \left(\frac{1}{k_{b2}^2 - \dot{e}_1^2} + 1 \right)(\tau(t) + EIy_{xx}(0,t)) \right)$$

$$+ \dot{e}_2 \left(\frac{e_2}{k_{b3}^2 - e_2^2} - (EIy_{xxx}(L,t) - k_2 e_2) + \left(\frac{1}{k_{b4}^2 - \dot{e}_2^2} + 1 \right) \right.$$

$$\left. (F(t) + EIy_{xxx}(L,t) - \alpha L(\tau(t) + EIy_{xx}(0,t)))) \right.$$

Substituting the boundary controllers (7.11) and (7.12) into above, we have

$$\dot{V}(t) = -k_3 \dot{e}_1^2 - k_4 \dot{e}_2^2 \leq 0 \qquad (7.15)$$

Considering the same PDE model, in this chapter, the dissipative and unique analysis are similar to the analysis of Chap. 8.

7.4 Convergence Analysis

It's obvious that $\dot{V}(t)$ is negative definite and $\dot{V}(t) \equiv 0$ when and only when $\dot{e}_1 \equiv \dot{e}_2 \equiv 0$ i.e. $\dot{\theta}(t) \equiv \dot{y}(L,t) \equiv 0$, further implying that $\ddot{\theta}(t) \equiv \ddot{y}(L,t) = 0$. Applying $\ddot{\theta}(t) = \ddot{y}(L,t) = 0$ into (7.1) and based on $z(x) = x\theta + y(x)$, we obtain

$$\rho \ddot{y}(x,t) = -EIy_{xxxx}(x,t) \qquad (7.16)$$

$$y_{xxxx}(L,t) = 0 \qquad (7.17)$$

Since (7.16) is a linear PDE with constant coefficients, we can derive the solutions by using the method of variables separation [3]:

$$y(x,t) = \varphi(x) \cdot \varsigma(t) \qquad (7.18)$$

where $\varphi(x)$ and $\varsigma(t)$ denote the unknown functions of space and time to be determined, respectively. With (7.16) and (7.18), it yields

$$\frac{\varphi''''(x)}{\varphi(x)} = -\frac{\rho}{EI} \frac{\varsigma''(t)}{\varsigma(t)} = \mu \qquad (7.19)$$

where $\varphi''''(x) = \frac{d^4 \varphi(x)}{dx^4}$, $\varsigma''(t) = \frac{d^2 \varsigma(t)}{dt^2}$.

Then we get

7.4 Convergence Analysis

$$\varphi''''(x) - \mu\varphi(x) = 0 \tag{7.20}$$

$$\varsigma''(t) + \frac{EI\mu}{\rho}\varsigma(t) = 0 \tag{7.21}$$

Setting $\mu = \beta^4$ and solving (7.20), we know

$$\varphi(x) = a_1 \cosh\beta x + a_2 \sinh\beta x + a_3 \cos\beta x + a_4 \sin\beta x \tag{7.22}$$

where $a_i \in R$, $i = 1, 2, 3, 4$ are pending positive constants. Based on (7.4), (7.17) and (7.18), we have $y(0,t) = y_x(0,t) = y_{xx}(L,t) = y_{xxxx}(L,t) = 0$. Then to solve (7.22), the following equations hold.

$$\begin{cases} a_1 + a_3 = 0 \\ a_2 + a_4 = 0 \\ a_1 \cosh\beta L + a_2 \sinh\beta L - a_3 \cos\beta L - a_4 \sin\beta L = 0 \\ a_1 \cosh\beta L + a_2 \sinh\beta L + a_3 \cos\beta L + a_4 \sin\beta L = 0 \end{cases} \tag{7.23}$$

It is quite clear that (7.23) have a unique solution $a_1 = a_2 = a_3 = a_4 = 0$. We further imply that $\varphi(x) = 0$ and $y(x,t) = 0$. It can easily show that $y_{xx}(0,t) = 0$ and $y_{xxx}(L,t) = 0$. Combining $y_{xx}(0,t) = 0$, $y_{xxx}(L,t) = 0$, (7.2) and (7.3) with (7.11) and (7.12), we have $e_1 = 0$ and $e_2 = 0$.

According to LaSalle's Invariance Principle [4], the closed-loop system is validated asymptotically stable under the controllers (7.11) and (7.12), i.e., $\theta(t) \to \theta_d$, $\dot{\theta}(t) \to \dot{\theta}_d$, $y(x,t) \to 0$, $y(L,t) \to y_d(L,t)$, $\dot{y}(L,t) \to \dot{y}_d(L,t)$, as $x \in [0,L]$, $t \to \infty$.

In addition, based on Lemma 7.1, the states $\theta(t)$, $\dot{\theta}(t)$, $y(L,t)$, $\dot{y}(L,t)$ for $t \in [0,\infty)$ are constrained.

7.5 Simulation Example

For the system (7.1)–(7.4), the physical parameters of the flexible manipulator are given as follows: $EI = 5.0$, $L = 1.0$, $\rho = 0.60$, $I_h = 0.50$, $m = 2.0$.

The desired signals of position and speed are $\theta_d = 0.3$, $\dot{\theta}_d = 0$, $y_d(L,t) = 0$, the system initial conditions are all set to be zeros, and the necessary parameters needed in the controllers are designed as: $k_{b1} = 0.4$, $k_{b2} = 0.2$, $k_{b3} = 0.2$, $k_{b4} = 0.2$, $k_1 = 200$, $k_2 = 200$, $k_3 = 200$, $k_4 = 200$. The performance of the whole closed-loop system is displayed by the simulation results as Figs. 7.3, 7.4, 7.5, 7.6 and 7.7.

Figures 7.3 and 7.4 show that the states $\theta(t)$, $\dot{\theta}(t)$, $y(L,t)$ and $\dot{y}(L,t)$ stay strictly within the sets $-0.1 < \theta(t) < 0.7$, $-0.2 < \dot{\theta}(t) < 0.2$, $-0.2 < y(L,t) < 0.2$, $-0.2 < \dot{y}(L,t) < 0.2$ under the proposed boundary control based on the Barrier Lyapunov function theory.

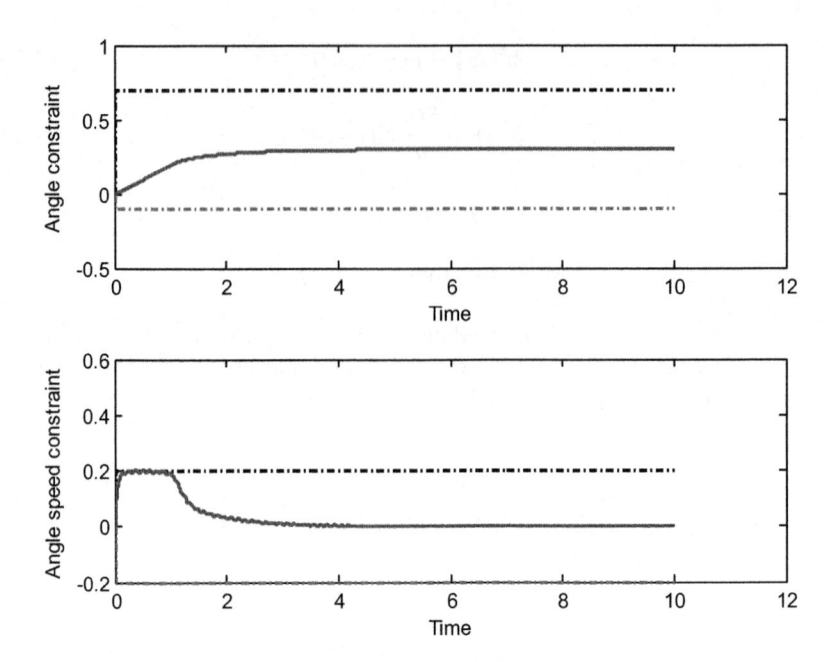

Fig. 7.3 Angle and its speed constraints

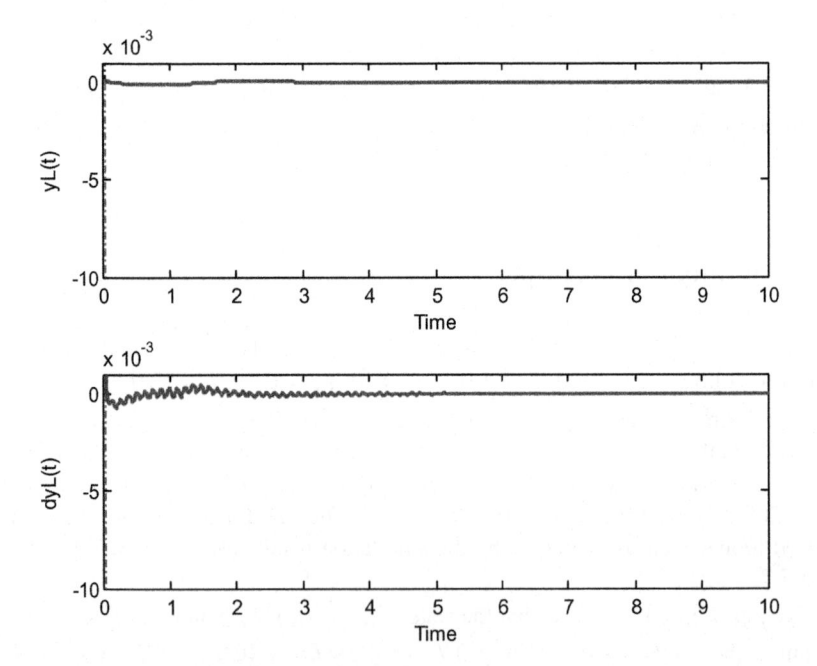

Fig. 7.4 Boundary deflection and its speed constraints

7.5 Simulation Example

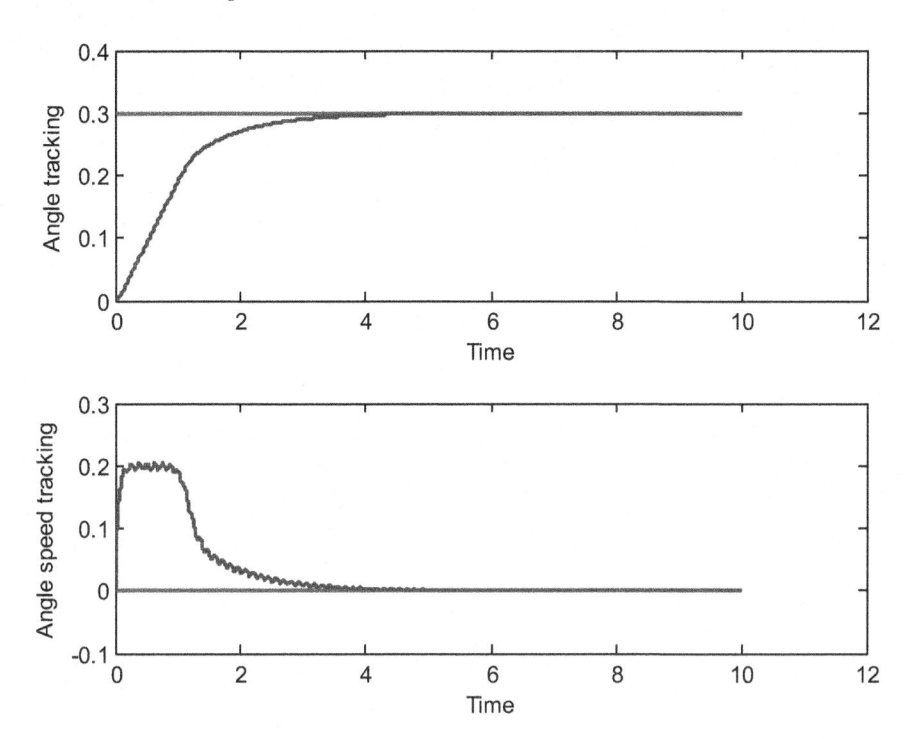

Fig. 7.5 Angle and angle speed response

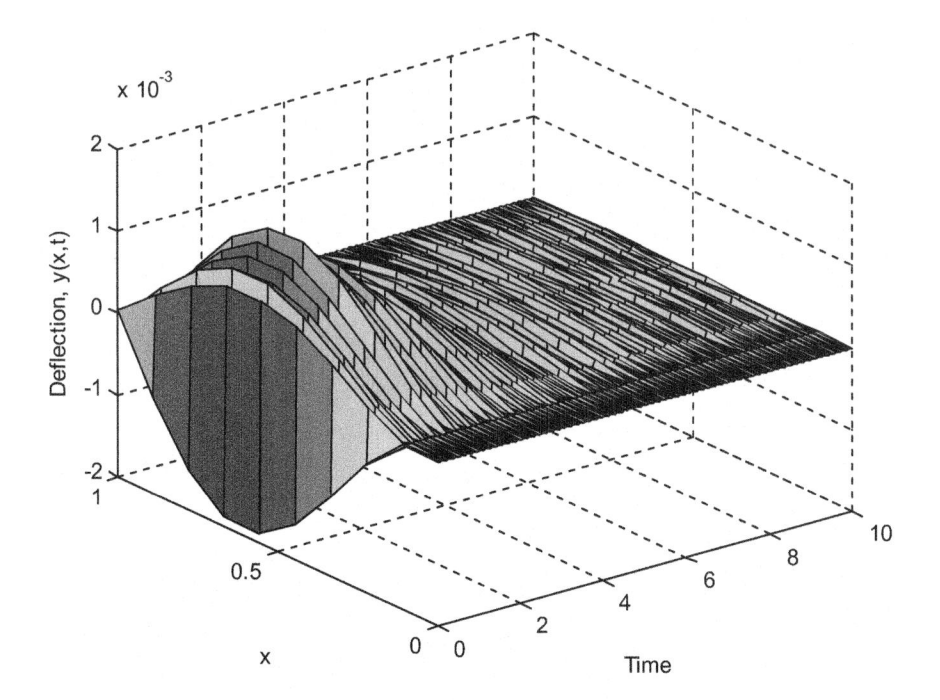

Fig. 7.6 Elastic deflection and its rate of the flexible manipulator

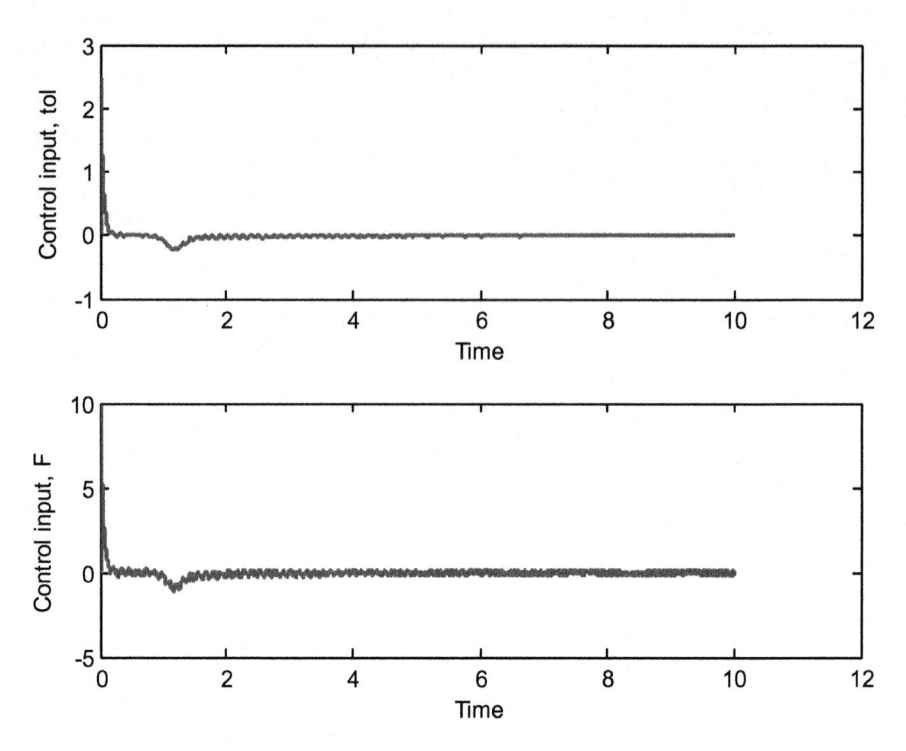

Fig. 7.7 Control inputs $\tau(t)$ and $F(t)$

7.5 Simulation Example

Simulation programs:

(1) Lyapunov function program: chap7_1.m

```
clear all;
close all;
ts=0.001;
kb=0.501;

for k=1:1:1001;
z(k)=(k-1)*ts-0.50;
V(k)=0.5*log(kb^2/(kb^2-z(k)^2));
end

figure(1);
plot(z,V,'r','linewidth',2);
xlabel('z');ylabel('V');
legend('Barrier Lyapunov function');
hold on;
plot(-kb,[0:0.001:3],'k',kb,[0:0.001:3],'k');
axis([-0.6 0.6 0 3]);
```

(2) Main program:chap7_2.m

```
close all;
clear all;
nx=10;
nt=20000;
```

```
tmax=10;L=1;
%Compute mesh spacing and time step
dx=L/(nx-1);
T=tmax/(nt-1);

%Create arrays to save data for export
t=linspace(0,nt*T,nt);
x=linspace(0,L,nx);

%Parameters
EI=5;rho=0.6;m=2;Ih=0.5;

k1=200;k2=200;k3=200;k4=200;

%Define variables and Initial condition:
y=zeros(nx,nt);
dy=zeros(nx,nt);

for j=1:nt
    th(j)=0;
    thd(j)=0.30;dthd(j)=0;ddthd(j)=0;

    tol(j)=0;F(j)=0;

    kb1(j)=0.4;
    kb2(j)=0.2;
    kb3(j)=0.2;
    kb4(j)=0.2;

    e2(j)=0;de2(j)=0;
    ydL(j)=0;dydL(j)=0;ddydL(j)=0;
end

for j=3:nt
alfa(j)=m/Ih;
yxx0(j-1)=(y(3,j-1)-2*y(2,j-1)+y(1,j-1))/dx^2;
yxx0(j)=(y(3,j)-2*y(2,j)+y(1,j))/dx^2;

th(j)=2*th(j-1)-th(j-2)+T^2/Ih*(tol(j-1)+EI*yxx0(j-1));
dth(j)=(th(j)-th(j-1))/T;
ddth(j)=(dth(j)-dth(j-1))/T;

e1(j)=th(j)-thd(j);
```

7.5 Simulation Example

```
de1(j)=dth(j)-dthd(j);

th_min(j)=-kb1(j)+thd(j);
th_max(j)=kb1(j)+thd(j);

dth_min(j)=-kb2(j)+dthd(j);
dth_max(j)=kb2(j)+dthd(j);
%get z(i,j),i=1,2, Boundary condition
y(1,:)=0;      %y(0,t)=0,  i=1
y(2,:)=0;      %y(1,t)=0,  i=2
z(1,:)=0;      %y(0,t)=0,  i=1
z(2,:)=0;      %y(1,t)=0,  i=2

%get y(i,j),i=3:nx-2
for i=3:nx-2

yxxxx(i,j-1)=(y(i+2,j-1)-4*y(i+1,j-1)+6*y(i,j-1)-4*y(i-1,j-1)+y(i-2,j
-1))/dx^4;

y(i,j)=T^2*(-i*dx*ddth(j-1)-(EI*yxxxx(i,j-1))/rho)+2*y(i,j-1)-y(i,j-2
);
   dy(i,j)=(y(i,j)-y(i,j-1))/T;
end

%get z(nx-1,j),i=nx-1
yxxxx(nx-1,j-1)=(-2*y(nx,j-1)+5*y(nx-1,j-1)-4*y(nx-2,j-1)+y(nx-3,j-1)
)/dx^4;
y(nx-1,j)=T^2*(-(nx-1)*dx*ddth(j-1)-(EI*yxxxx(nx-1,j-1))/rho)+2*y(nx-
1,j-1)-y(nx-1,j-2);    %(19)
dy(nx-1,j)=(y(nx-1,j)-y(nx-1,j-1))/T;
%get y(nx,j),y=nx
yxxxL(j-1)=(-y(nx,j-1)+2*y(nx-1,j-1)-y(nx-2,j-1))/dx^3;

F(j-1)=-EI*yxxxL(j-1)+alfa(j-1)*L*(tol(j-1)+EI*yxx0(j-1))+((kb4(j-1)^
2-de2(j-1)^2)/(kb4(j-1)^2-de2(j-1)^2+1)*(-e2(j-1)/(kb3(j-1)^2-e2(j-1)
^2)+EI*yxxxL(j-1)-k4*e2(j-1)-k2*de2(j-1)));
y(nx,j)=T^2*(-L*ddth(j-1)+(EI*yxxxL(j-1)+F(j-1))/m)+2*y(nx,j-1)-y(nx,
j-2);

yxxxL(j)=(-y(nx,j)+2*y(nx-1,j)-y(nx-2,j))/dx^3;

yL(j)=y(nx,j);
dyL(j)=(yL(j)-yL(j-1))/T;
```

```
dy(nx,j)=(y(nx,j)-y(nx,j-1))/T;

yL_min(j)=-kb3(j)+ydL(j);
yL_max(j)=kb3(j)+ydL(j);

dyL_min(j)=-kb4(j)+dydL(j);
dyL_max(j)=kb4(j)+dydL(j);

e2(j)=yL(j)-ydL(j);
de2(j)=dyL(j)-dydL(j);

tol(j)=-EI*yxx0(j)+((kb2(j)^2-de1(j)^2)/(kb2(j)^2-de1(j)^2+1))*(-e1(j
)/(kb1(j)^2-e1(j)^2)+EI*L*yxxxL(j)+EI*yxx0(j)-k1*e1(j)-k3*de1(j));
end
%To view the curve, short the points

tshort=linspace(0,tmax,nt/100);
yshort=zeros(nx,nt/100);
dyshort=zeros(nx,nt/100);
for j=1:nt/100
    for i=1:nx
        yshort(i,j)=y(i,j*100);    %Using true y(i,j)
        dyshort(i,j)=dy(i,j*100);
    end
end

figure(1);
subplot(211);
plot(t,th_min,'-.r',t,th_max,'-.k',t,th,'b','linewidth',2);
xlabel('Time');ylabel('Angle constraint');
subplot(212);
plot(t,dth_min,'-.r',t,dth_max,'-.k',t,dth,'b','linewidth',2);
xlabel('Time');ylabel('Angle speed constraint');

figure(2);
subplot(211);
plot(t,yL_min,'-.r',t,yL_max,'-.k',t,yL,'b','linewidth',2);
xlabel('Time');ylabel('yL(t)');
axis([0 10 -0.005 0.005]);
subplot(212);
plot(t,dyL_min,'-.r',t,dyL_max,'-.k',t,dyL,'b','linewidth',2);
xlabel('Time');ylabel('dyL(t)');
axis([0 10 -0.005 0.005]);
```

```
figure(3);
subplot(211);
plot(t,thd,'r',t,th,'b','linewidth',2);
xlabel('Time');ylabel('Angle tracking');
subplot(212);
plot(t,dthd,'r',t,dth,'b','linewidth',2);
xlabel('Time');ylabel('Angle speed tracking');

figure(4);
surf(tshort,x,yshort);
xlabel('Time'); ylabel('x');zlabel('Deflection, y(x,t)');

figure(5);
subplot(211);
plot(t,tol,'r','linewidth',2);
xlabel('Time');ylabel('Control input, tol');
subplot(212);
plot(t,F,'r','linewidth',2);
xlabel('Time');ylabel('Control input, F');
```

References

1. T.T. Jiang, J.K. Liu, W. He, Adaptive boundary control for a flexible manipulator with state constraints using a barrier lyapunov function, J. Dyn. Syst. Meas. Control (2017) (accepted)
2. K.P. Tee, S.S. Ge, E.H. Tay, Barrier Lyapunov functions for the control of output-constrained nonlinear systems. Automatica **45**, 918–927 (2009)
3. W.H. Ray, *Advanced Process Control* (McGraw-Hill, New York, 1981)
4. C.D. Rahn, *Mechatronic Control of Distributed Noise and Vibration-A Lyapunov Approach* (Springer, Berlin Heidelberg, 2001)

Chapter 8
Boundary Control of Flexible Manipulator with Input Constraints

8.1 Introduction

Previous studies have considered the stability problem under the condition of input constraints [1], which are based on nested saturated input functions. Some researchers present anti-windup controllers for the linear system using linear matrix inequalities [2]. In [3], a simple controller with the smooth hyperbolic function for achieving trajectory tracking under the condition of restricted input is presented, and this idea is helpful for our work. However, the PDE model is so complex that it needs efforts to reduce analytical complexity and design the control scheme with input constraints.

In the last several years, despite the significant progress of the control design for flexible manipulator systems, studies of flexible manipulators or PDEs with input constraints are limited. In this chapter, based on the work [4], we consider the trajectory tracking and vibration suppression control problem in a flexible manipulator with restricted inputs.

8.2 System Description

Considering PDE model given in Chap. 3, neglecting disturbance, the PDE model is given as

$$\rho\left(x\ddot{\theta} + \ddot{y}(x)\right) = -EIy_{xxxx}(x) \tag{8.1}$$

$$\tau = I_h\ddot{\theta} - EIy_{xx}(0) \tag{8.2}$$

$$F = m\left(L\ddot{\theta} + \ddot{y}(L)\right) - EIy_{xxx}(L) \tag{8.3}$$

$$y(0) = 0, \ y_x(0) = 0, \ y_{xx}(L) = 0 \tag{8.4}$$

Define $z(x) = x\theta + y(x)$, then $\ddot{z}(x) = x\ddot{\theta} + \ddot{y}(x)$, $\ddot{z}(L) = x\ddot{\theta} + \ddot{y}(L)$. From $z(x)$ definition, we have

$$z_{xx}(x) = y_{xx}(x), \ddot{z}_x(0) = \ddot{\theta}, z_{xx}(0) = y_{xx}(0), z_{xx}(L) = y_{xx}(L), z_{xxx}(L) = y_{xxx}(L)$$

Smooth hyperbolic tangent function is

$$\tanh(x) = \frac{e^x - e^{-x}}{e^x + e^{-x}}$$

The function have following properties: (1) $|\tanh(x)| \leq 1$; (2); $x \tanh(x) \geq 0$.

Consider $\theta_d(t)$ as an constant ideal angle signal, the control goals are: $\theta(t) \to \theta_d(t)$, $\dot{\theta}(t) \to 0$, $y(x,t) \to 0$ under input constraints.

Define θ_d as constant value, then we have $e = \theta - \theta_d$, $\dot{e} = \dot{\theta} - \dot{\theta}_d = \dot{\theta}$, $\ddot{e} = \ddot{\theta} - \ddot{\theta}_d = \ddot{\theta}$.

8.3 Controller Design

Design the inputs u and F as

$$\tau = -\alpha_1 l_1 \tanh(l_1 e) - \alpha_2 l_2 \tanh(l_2 \dot{e}) \tag{8.5}$$

$$F = -\alpha_3 l_3 \tanh(l_3 \dot{z}(L, t)) \tag{8.6}$$

where $\alpha_1, l_1, \alpha_2, l_2, \alpha_3, l_3 > 0$.

Theorem 8.1 [2] The controller (8.5) and (8.6) can guarantee the asymptotical stability of the system, i.e., $\theta \to \theta_d$ and $y(x,t) \to 0 \, \forall x \in [0, L]$ when $t \to +\infty$.

Proof The Lyapunov candidate function is taken to be

$$V = E_1 + E_2 \tag{8.7}$$

where $E_1 = \frac{1}{2}\int_0^L \rho \dot{z}^2(x,t)dx + \frac{1}{2}EI \int_0^L y_{xx}^2(x,t)dx$, $E_2 = \frac{1}{2}I_h \dot{e}^2 + \frac{1}{2}m\dot{z}^2(L,t) + \alpha_1 \ln(\cosh(l_1 e))$, $\cosh(x) \in [1, +\infty)$, $\ln(\cosh(l_1 e)) \geq 0$.

In above Lyapunov function, the manipulator kinetic energy $\frac{1}{2}\int_0^L \rho \dot{z}^2(x,t)dx$, the manipulator potential energy $\frac{1}{2}EI \int_0^L y_{xx}^2(x,t)dx$, and the load kinetic energy are all considered.

8.3 Controller Design

Then

$$\dot{V} = \dot{E}_1 + \dot{E}_2$$

where

$$\begin{aligned}
\dot{E}_1 &= \int_0^L \rho \dot{z}(x,t) \ddot{z}(x,t) dx + EI \int_0^L y_{xx}(x,t) \dot{y}_{xx}(x,t) dx \\
&= -EI y_{xxx}(L,t) L \dot{\theta}(t) - EI y_{xx}(0,t) \dot{\theta}(t) - EI y_{xxx}(L,t) \dot{y}(L,t) \\
&= -EI y_{xxx}(L,t) \dot{z}(L,t) - EI y_{xx}(0,t) \dot{\theta}(t)
\end{aligned}$$

$$\begin{aligned}
\dot{E}_2 &= I_h \dot{e} \ddot{e} + m \dot{z}(L,t) \ddot{z}(L,t) + \alpha_1 l_1 \dot{e} \tanh(l_1 e) \\
&= \dot{e}(I_h \ddot{e} + \alpha_1 l_1 \tanh(l_1 e)) + \dot{z}(L,t) m \ddot{z}(L,t)
\end{aligned}$$

It follows that

$$\begin{aligned}
\dot{V} = \dot{E}_1 + \dot{E}_2 &= -EI y_{xxx}(L,t) \dot{z}(L,t) - EI y_{xx}(0,t) \dot{\theta}(t) + \dot{e}(I_h \ddot{e} + \alpha_1 l_1 \tanh(l_1 e)) + \dot{z}(L,t) m \ddot{z}(L,t) \\
&= \dot{e}(I_h \ddot{e} + \alpha_1 l_1 \tanh(l_1 e) - EI y_{xx}(0,t)) + \dot{z}(L,t)(-EI y_{xxx}(L,t) + m \ddot{z}(L,t)) \\
&= \dot{e}(\tau + \alpha_1 l_1 \tanh(l_1 e)) + \dot{z}(L,t) F
\end{aligned}$$

Substituting (8.5) and (8.6) into above yields

$$\begin{aligned}
\dot{V} &= \dot{e}(\tau + \alpha_1 l_1 \tanh(l_1 e)) + \dot{z}(L,t) F \\
&= -\alpha_2 l_2 \dot{e} \tanh(l_2 \dot{e}) - \alpha_3 l_3 \dot{z}(L,t) \tanh(l_3 \dot{z}(L,t)) \leq 0
\end{aligned} \qquad (8.8)$$

8.4 Dissipative Analysis of the Closed System

Define

$$\mathbf{q} = [q_1 \quad q_2 \quad q_3 \quad q_4 \quad q_5]^T = [e \quad \dot{e} \quad \dot{z}(L) \quad y(x) \quad \dot{y}(x)]^T \qquad (8.9)$$

The closed-loop system can be compactly written as

$$\dot{\mathbf{q}} = \mathcal{A}\mathbf{q}, \quad \mathbf{q}(0) \in \mathcal{H}$$

The spaces related to $\mathbf{q}$ mentioned above are defined as

$$\mathcal{H} = \mathbb{R}^3 \times H^2 \times L^2 \qquad (8.10)$$

where $L^2(\Omega) = \left\{ f \mid \int_\Omega |f(x)|^2 dx < \infty \right\}$,

$H^k(\Omega) = \left\{ f | f, f', \ldots, f^{(k-1)} \text{are absolutely continuous}, f^{(k)} \in L^2(\Omega) \right\}, \Omega = [0, L].$

$H^k(\Omega) = \left\{ f | f, f', \ldots, f^{(k-1)} \text{are absolutely continuous}, f^{(k)} \in L^2(\Omega) \right\}$

In $\mathcal{H}$, $\mathbb{R}^3$ is defined for q_1, that is, q_1, q_2 and q_3 are all real number. H^2 is defined for q_4, that is, $y(x, t)$, $y_x(x, t)$ and $y_{xx}(x, t)$ are all L^2 limited, L^2 is defined for q_5, that is, $\dot{y}(x, t)$ is L_2 limited.

$\mathcal{A}$ is a infinite dimensional linear operator, define

$$\mathcal{A}q = [\dot{q}_1 \; \dot{q}_2 \; \dot{q}_3 \; \dot{q}_4 \; \dot{q}_5]^T, \quad \forall q \in \mathcal{D}(\mathcal{A}) \tag{8.11}$$

From (8.2), we have $\tau = I_h \ddot{\theta} - EI y_{xx}(0)$, then we get

$$\ddot{e} = \frac{1}{I_h}(\tau + EI y_{xx}(0)) = \frac{1}{I_h}(-\alpha_1 l_1 \tanh(l_1 e) - \alpha_2 l_2 \tanh(l_2 \dot{e}) + EI y_{xx}(0))$$

Considering $y_{xx}(0) = q_{4,xx}(0)$, then we get

$$\dot{q}_2 = \frac{1}{I_h}(-\alpha_1 l_1 \tanh(l_1 q_1) - \alpha_2 l_2 \tanh(l_2 q_2) + EI q_{4,xx}(0)) \tag{8.12}$$

From $F = m\left(L\ddot{\theta} + \ddot{y}(L)\right) - EI y_{xxx}(L)$, consider $y_{xxx}(L, t) = q_{4,xxx}(L)$ and $q_3 = \dot{z}(L, t)$, we have $\dot{q}_3 = \ddot{z}(L, t) = \frac{1}{m}(EI y_{xxx}(L, t) + F) = \frac{1}{m}\left(EI q_{4,xxx}(L) - \alpha_3 l_3 \tanh(l_3 \dot{z}(L, t))\right)$. Then we get

$$\dot{q}_3 = \frac{1}{m}\left(EI q_{4,xxx}(L) - \alpha_3 l_3 \tanh(l_3 q_3)\right) \tag{8.13}$$

From $\rho\left(x\ddot{\theta}(t) + \ddot{y}(x, t)\right) = -EI y_{xxxx}(x, t)$, we have

$$\ddot{y}(x, t) = -\frac{1}{\rho} EI y_{xxxx}(x) - x\ddot{\theta}(t)$$

Submitting $\ddot{\theta} = \frac{1}{I_h}(-\alpha_1 l_1 \tanh(l_1 e) - \alpha_2 l_2 \tanh(l_2 \dot{e}) + EI y_{xx}(0))$ into above, considering $q_4 = y(x, t)$,

we get

$$\ddot{y}(x, t) = -\frac{1}{\rho} EI q_{4,xxxx}(x) - \frac{x}{I_h}(-\alpha_1 l_1 \tanh(l_1 q_1) - \alpha_2 l_2 \tanh(l_2 q_2) + EI q_{4,xx}(0))$$

8.4 Dissipative Analysis of the Closed System

Then we have

$$\mathcal{A}q = \begin{bmatrix} q_2 \\ \frac{1}{I_h}\left(-\alpha_1 l_1 \tanh(l_1 q_1) - \alpha_2 l_2 \tanh(l_2 q_2) + EIq_{4,xx}(0)\right) \\ \frac{1}{m}\left(EIq_{4,xxx}(L) - \alpha_3 l_3 \tanh(l_3 q_3)\right) \\ q_5 \\ -\frac{1}{\rho} EIq_{4,xxxx}(x) - \frac{x}{I_h}\left(-\alpha_1 l_1 \tanh(l_1 q_1) - \alpha_2 l_2 \tanh(l_2 q_2) + EIq_{4,xx}(0)\right) \end{bmatrix}$$
(8.14)

In $\mathcal{H}$ and $\mathcal{D}(\mathcal{A})$, since $\dot{V} \leq 0$, the operator $\mathcal{A}$ is dissipative.

8.5 Unique Analysis of Solutions

To prove the unique solutions of q and $\mathcal{A}^{-1}$ is a compact operator, define $g = \begin{bmatrix} g_1 & g_2 & g_3 & g_4 & g_5 \end{bmatrix}^T \in \mathcal{H}$, let

$$\mathcal{A}q = g \tag{8.15}$$

From (8.14), we have

$$g_1 = q_2$$
$$g_2 = \frac{1}{I_h}\left(-\alpha_1 l_1 \tanh(l_1 q_1) - \alpha_2 l_2 \tanh(l_2 q_2) + EIq_{4,xx}(0)\right)$$
$$g_3 = \frac{1}{m}\left(EIq_{4,xxx}(L) - \alpha_3 l_3 \tanh(l_3 q_3)\right)$$
$$g_4 = q_5$$
$$g_5 = -\frac{1}{\rho} EIq_{4,xxxx}(x) - \frac{x}{I_h}\left(-\alpha_1 l_1 \tanh(l_1 q_1) - \alpha_2 l_2 \tanh(l_2 q_2) + EIq_{4,xx}(0)\right)$$
(8.16)

From g_2 and g_5 expression, we have $g_5 = -\frac{1}{\rho} EIq_{4,xxxx}(x) - xg_2$, i.e.

$$q_{4,xxxx}(x) = \frac{\rho}{EI}(-g_5 - xg_2)$$

The solution is

$$q_4 = -\frac{\rho}{EI}\int_0^x \int_0^{\xi_1} \int_0^{\xi_2} \int_0^{\xi_3} g_5(\xi_4) d\xi_4 d\xi_3 d\xi_2 d\xi_1 - \frac{x^5}{5!}\frac{\rho}{EI_h}g_2 + \sum_{j=0}^{3} \sigma_j x^j$$

where $\sigma_0, \ldots, \sigma_3$ are uniquely determined by boundary conditions (8.3) and (8.4).

From g_2 expression, we have

$$q_1 = \frac{1}{l_1}\operatorname{arctanh}\left(EIq_{4,xx}(0) - I_h g_2 - \alpha_2 l_2 \tanh(l_2 q_2)\right)/\alpha_1 l_1$$

From $z(x) = x\theta + y(x)$, we have $\dot{z}(L) = L\dot{\theta} + \dot{y}(L)$, from (8.9). Then we have

$$q_3 = q_4(L) + Lq_2$$

Then we can get the unique solution as

$q_1 = \dfrac{1}{l_1}\operatorname{arctanh}\left(EIq_{4,xx}(0) - I_h g_2 - \alpha_2 l_2 \tanh(l_2 q_2)\right)/\alpha_1 l_1$

$q_2 = g_1$

$q_3 = q_4(L) + Lq_2$

$q_4 = -\dfrac{\rho}{EI}\displaystyle\int_0^x \int_0^{\xi_1} \int_0^{\xi_2} \int_0^{\xi_3} g_5(\xi_4)\,d\xi_4 d\xi_3 d\xi_2 d\xi_1 - \dfrac{x^5}{5!}\dfrac{\rho}{EI}g_2 + \displaystyle\sum_{j=0}^{3}\sigma_j x^j$ \hfill (8.17)

$q_5 = g_4$

Hence, Equation $\mathcal{A}q = g$ has a unique solution $q \in \mathcal{D}(\mathcal{A})$, implying that $\mathcal{A}^{-1}$ exists and maps $\mathcal{H}$ into $\tilde{\mathcal{H}} = \mathbb{R}^3 \times H^2 \times L^2$. Moreover, since $\mathcal{A}^{-1}$ maps every bounded set of $\mathcal{H}$ into bounded set of $\tilde{\mathcal{H}} = \mathbb{R}^3 \times H^2 \times L^2$, the embedding of the later space onto $\mathcal{H}$ is compact. It follows that $\mathcal{A}^{-1}$ is a compact operator.

The spectrum of $\mathcal{A}$ consists entirely of isolated eigenvalues. It also proves that for any $\lambda > 0$ in the resolvent set of $\mathcal{A}$, the operator $(\lambda I - \mathcal{A})^{-1}$ is a compact operator. Based on the Lumer-Phillips theorem, operator $\mathcal{A}$ generates a C_0-semi-group of contractions $T(t)$ on $\mathcal{H}$ [4].

8.6 Convergence Analysis

To apply the extended LaSalle's invariance principle, we need prove to show that $\dot{V} \equiv 0$ implies $y(x,t) \equiv 0$.

If $\dot{V} \equiv 0$, then

$$\dot{e} \equiv \dot{z}(L,t) \equiv 0 \tag{8.18}$$

8.6 Convergence Analysis

Further implying that $\ddot{e} \equiv \ddot{z}(L,t) \equiv 0$. From (8.1), we have

$$\rho \ddot{z}(L,t) = -EI y_{xxxx}(L,t)$$

then we get

$$y_{xxxx}(L,t) = 0 \qquad (8.19)$$

Note that Eq. (8.1) is separable and can be treated by the technique of separation of variables [5], then we write $y(x,t)$ as follows:

$$y(x,t) = W(x) \cdot \phi(t) \qquad (8.20)$$

where $W(x)$ and $\phi(t)$ are unknown functions of space and time to be determined. Considering $\ddot{\theta} \equiv 0$, from (8.1), we have $\rho \ddot{y}(x) = -EI y_{xxxx}(x)$, then

$$y_{xxxx}(x,t) = -\frac{\rho}{EI} \ddot{y}(x,t)$$

From (8.19), we have $y_{xxxx}(x,t) = W^{(4)}(x) \cdot \phi(t)$, $\ddot{y}(x,t) = W(x) \cdot \phi''(t)$, then above equation becomes

$$\frac{W^{(4)}(x)}{W(x)} = -\frac{\rho}{EI} \frac{\phi''(t)}{\phi(t)} = \mu$$

where $\phi''(t) = \frac{d^2\phi(t)}{dt^2}$, $W^{(4)}(x) = \frac{d^4 W}{dx^4}$.

Then we have

$$W^{(4)}(x) - \mu W(x) = 0 \qquad (8.21)$$

$$\phi''(t) + \frac{EI\mu}{\rho} \phi(t) = 0 \qquad (8.22)$$

Let $\mu = \eta^4$, then we can get the solution (8.19) as

$$W(x) = c_1 \cosh \eta x + c_2 \sinh \eta x + c_3 \cos \eta x + c_4 \sin \eta x \qquad (8.23)$$

where $c_i \in R$, $i = 1, 2, 3, 4$ are unknown real number to be determined.

Considering (8.4) and (8.18), we have $W(0) = W'(0) = W''(L) = W^{(4)}(L) = 0$, then from (8.21), we have

$$\begin{cases} c_1 + c_3 = 0 \\ c_2 + c_4 = 0 \\ c_1 \cosh \beta L + c_2 \sinh \beta L - c_3 \cos \beta L - c_4 \sin \beta L = 0 \\ c_1 \cosh \beta L + c_2 \sinh \beta L + c_3 \cos \beta L + c_4 \sin \beta L = 0 \end{cases} \qquad (8.24)$$

Then we have

$$\begin{cases} c_1 \cosh \beta L + c_2 \sinh \beta L = 0 \\ c_3 \cos \beta L + c_4 \sin \beta L = 0 \end{cases}$$

i.e.

$$\begin{cases} c_3 \cosh \beta L + c_4 \sinh \beta L = 0 \\ c_3 \cos \beta L + c_4 \sin \beta L = 0 \end{cases}$$

therefore

$$c_4(\sinh \beta L \cdot \cos \beta L - \sin \beta L \cdot \cosh \beta L) = 0$$

We can conclude that $W^{(4)}(x) - \mu W(x) = 0$ have unique solutions, $c_i = 0$, $i = 1, 2, 3, 4$, thus, $W(x) = 0$, $y(x,t) = 0$ and $W_{xx}(0) = -c_1 + c_3 = 0$. Consider $y(x,t) = W(x) \cdot \phi(t)$, we have $y_{xx}(0) = W_{xx}(0) \cdot \phi(t) = 0$. Substitute $\tau = -\alpha_1 l_1 \tanh(l_1 e) - \alpha_2 l_2 \tanh(l_2 \dot{e})$ into (8.2), we have

$$-\alpha_1 l_1 \tanh(l_1 e) - \alpha_2 l_2 \tanh(l_2 \dot{e}) = I_h \ddot{\theta} - EI y_{xx}(0)$$

Considering if $\dot{V} \equiv 0$, then $\dot{e} \equiv 0$, $\ddot{\theta} \equiv 0$, combining with $y_{xx}(0) = 0$, then we have $e = 0$..

Therefore, according to the extended LaSalle's invariance principle [6], the PDE boundary control (8.5) and (8.6) can guarantee the asymptotic stability of the closed-loop system. If $t \to \infty$, $e \to 0$, $\dot{e} \to 0$, $y(x,t) \to 0$.

Considering the boundary control (8.5) and (8.6) and noting $\tanh(x) \in [-1, +1]$, it follows that

$$|\tau| = |-\alpha_1 l_1 \tanh(l_1 e) - \alpha_2 l_2 \tanh(l_2 \dot{e})| \le \alpha_1 l_1 + \alpha_2 l_2 = u_{\max} \quad (8.25)$$

$$|F| = |-\alpha_3 l_3 \tanh(l_3 \dot{z}(L,t))| \le \alpha_3 l_3 = F_{\max} \quad (8.26)$$

So we can set parameters $\alpha_1, \alpha_2, \alpha_3, l_1, l_2$ and l_3 to adjust limits $u_{\max}$ and $F_{\max}$.

8.7 Simulation Example

Considering the PDE model as Eqs. (8.1)–(8.4), the physical parameters are chosen as: $EI = 2.0$, $L = 1.0$, $\rho = 0.20$, $m = 0.20$, $I_h = 0.50$.

Define ideal angle as $\theta_d = 0.5$. Use controller (8.5) and (8.6). Let $\alpha_1 = 5$, $\alpha_2 = 5$, $\alpha_3 = 8$, $l_1 = l_2 = l_3 = 0.12$, $\alpha_2 = 5$, $\alpha_3 = 8$. Two axes are divided into sections as $nx = 9$, $nt = 20001$. The simulation results are shown from Figs. 8.1, 8.2, 8.3 and 8.4.

8.7 Simulation Example

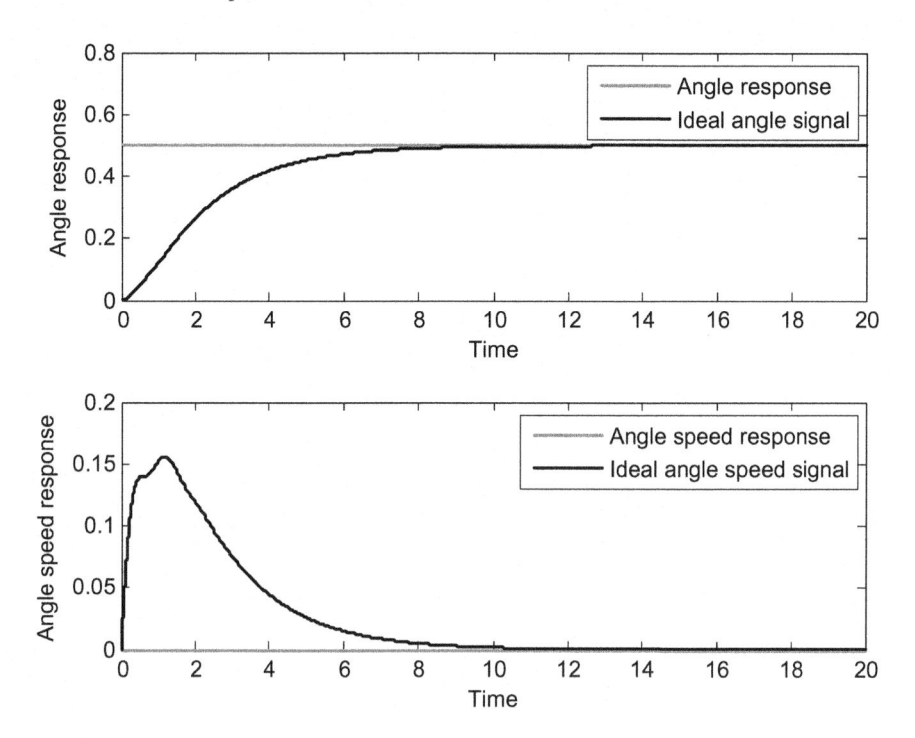

Fig. 8.1 Angle tracking and angle speed tracking

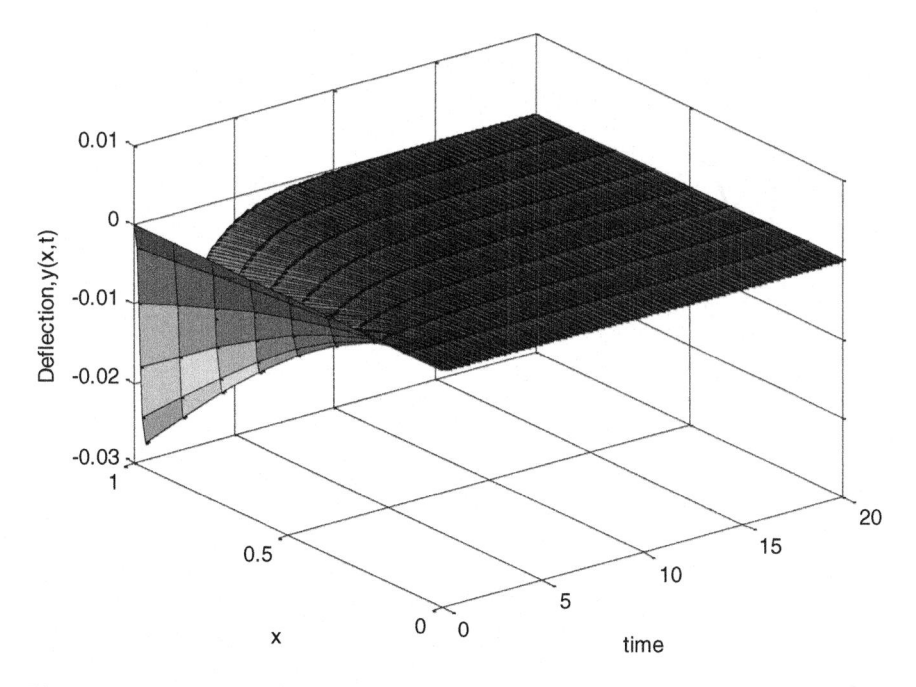

Fig. 8.2 Deformation and deformation rate

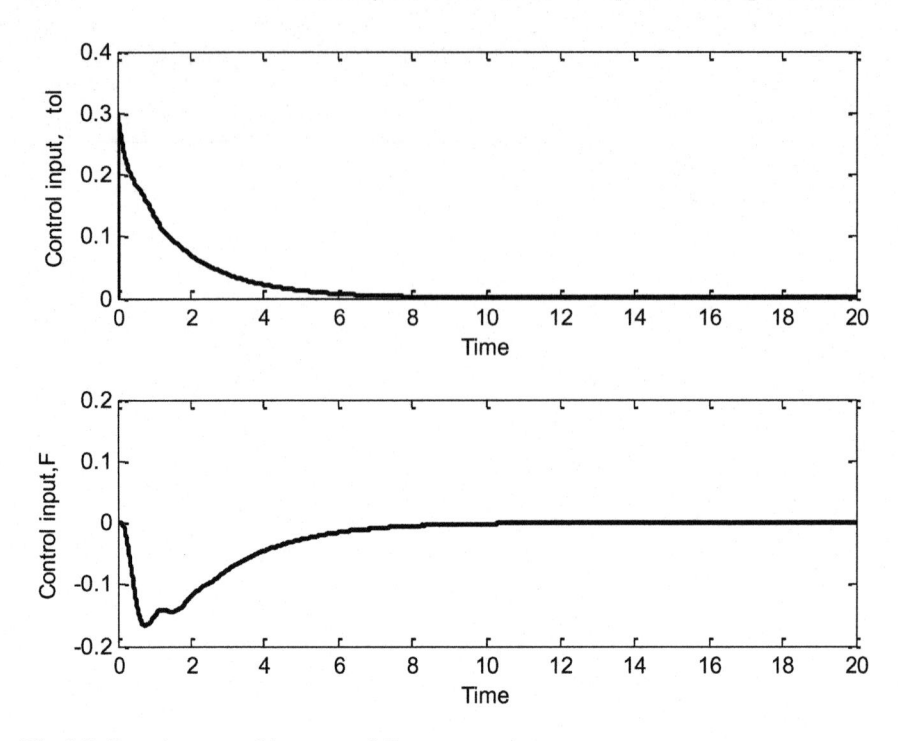

Fig. 8.3 Boundary control input, τ and F

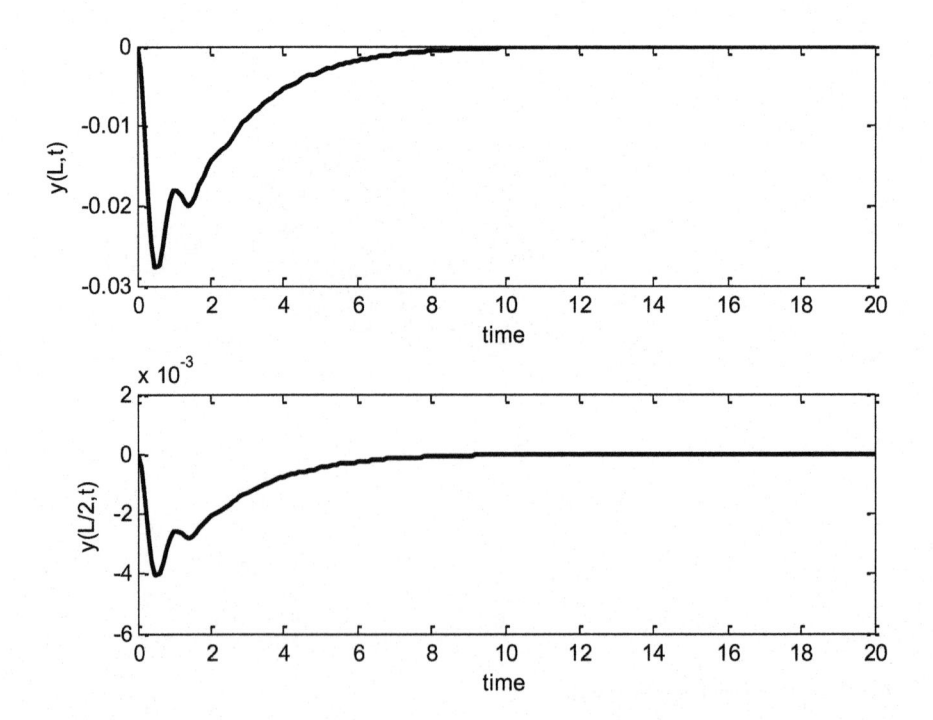

Fig. 8.4 Deformation at $x = \frac{L}{2}$ and $x = L$

8.7 Simulation Example

Simulation program: chap8_1.m

```
close all;
clear all;

nx=8+1;nt=20000+1;
tmax=20.0;L=1;
%Compute mesh spacing and time step
dx=L/(nx-1);
T=tmax/(nt-1);

%Create arrays to save data for export
t=linspace(0,tmax,nt);
x=linspace(0,L,nx);
%Parameters
EI=2;rho=0.2;m=0.2;Ih=0.5;

dyLj=0;
%Define viriables and Initial condition:
y=zeros(nx,nt);    %elastic deflectgion
th_2=0;th_1=0;
for j=1:nt
    th(j)=0;    %joint angle
    thd(j)=0.5;
    dthd(j)=0;
end

for j=3:nt      %Begin

dth(j)=(th_1-th_2)/T;
alfa1=5;alfa2=5;alfa3=8;
l1=0.12;l2=0.12;l3=0.12;

e(j)=th_1-thd(j);
de(j)=dth(j)-dthd(j);
e1(j)=e(j);
e2(j)=de(j);

tol(j)=-1*alfa1*tanh(l1*e(j))-alfa2*tanh(l2*de(j));

dzL(j)=L*dth(j)+dyLj;
```

```
F(j)=-alfa3*tanh(l3*dzL(j));

yxx0=(y(3,j-1)-2*y(2,j-1)+y(1,j-1))/dx^2;
th(j)=2*th_1-th_2+T^2/Ih*(tol(j)+EI*yxx0);
ddth(j)=(th(j)-2*th_1+th_2)/T^2;

%get y(i,j),i=1,2, Boundary condition
y(1,:)=0;     %y(0,t)=0, i=1
y(2,:)=0;     %y(1,t)=0, i=2

%get y(i,j),i=3:nx-2
for i=3:nx-2
    yxxxx=(y(i+2,j-1)-4*y(i+1,j-1)+6*y(i,j-1)-4*y(i-1,j-1)+y(i-2,j-1))/dx^4;
    y(i,j)=T^2*(-i*dx*ddth(j)-EI/rho*yxxxx)+2*y(i,j-1)-y(i,j-2);  %i*dx=x
end

%get y(nx-1,j),i=nx-1
yxxxx(nx-1,j-1)=(-2*y(nx,j-1)+5*y(nx-1,j-1)-4*y(nx-2,j-1)+y(nx-3,j-1))/dx^4; %yxx
xx(nx-1,j-1)
y(nx-1,j)=T^2*(-(nx-1)*dx*ddth(j)-EI/rho*yxxxx(nx-1,j-1))+2*y(nx-1,j-1)-y(nx-1,j-
2);

%get y(nx,j),y=nx
yxxx_L=(-y(nx,j-1)+2*y(nx-1,j-1)-y(nx-2,j-1))/dx^3;
y(nx,j)=T^2*(-L*ddth(j)+(EI*yxxx_L+F(j))/m)+2*y(nx,j-1)-y(nx,j-2);
dyL(j)=(y(nx,j-1)-y(nx,j-2))/T;

th_2=th_1;
th_1=th(j);
dyLj=dyL(j);
end   %End

%To view the curve, short the points
tshort=linspace(0,tmax,(nt-1)/100+1);
yshort=zeros(nx-1,(nt-1)/100+1);
for j=1:(nt-1)/100+1
    for i=1:nx
        yshort(i,j)=y(i,(j-1)*100+1);
    end
end
figure(1);
subplot(2,1,1)
```

8.7 Simulation Example

```
plot(t,thd,'r',t,th,'k','linewidth',2);
legend('Angle response','Ideal angle signal');
xlabel('Time');ylabel('Angle response');
subplot(2,1,2)
plot(t,dthd,'r',t,dth,'k','linewidth',2);
legend('Angle speed response','Ideal angle speed signal');
xlabel('Time');ylabel('Angle speed response');

figure(2);
surf(tshort,x,yshort);
colormap(summer);
xlabel('time'); ylabel('x');zlabel('Deflection,y(x,t)');
colormap('Jet')

figure(3);
subplot(2,1,1)
plot(t,tol,'k','linewidth',2);
xlabel('Time');ylabel('Control input£¬tol');
subplot(2,1,2)
plot(t,F,'k','linewidth',2);
xlabel('Time');ylabel('Control input,F');

figure(4);
subplot(211);
for j=1:(nt-1)/100+1
    yshortL(j)=y(nx,(j-1)*100+1);
end
plot(tshort,yshortL,'k','linewidth',2);
xlabel('time');ylabel('y(L,t)');
subplot(212);
for j=1:(nt-1)/100+1
    yshort1(j)=y((nx-1)/2,(j-1)*100+1);
end
plot(tshort,yshort1,'k','linewidth',2);
xlabel('time');ylabel('y(L/2,t)');
```

References

1. C. Wen, J. Zhou, Z. Liu, H. Su, Robust adaptive control of uncertain nonlinear systems in the presence of input saturation and external disturbance. IEEE Trans. Autom. Control **56**(7), 1672–1678 (2011)
2. E.F. Mulder, M.V. Kothare, M. Morari, Multivariable anti-windup controller synthesis using linear matrix inequalities. Automatica **37**(9), 407–1416 (2001)
3. A. Ailon, Simple tracking controllers for autonomous VTOL aircraft with bounded inputs. IEEE Trans. Autom. Control **55**(3), 737–743 (2010)
4. Z.J. Liu, J.K. Liu, Partial differential equation boundary control of a flexible manipulator with input saturation. Int. J. Syst. Sci. **48**(1), 53–62 (2017)
5. W.H. Ray, *Advanced Process Control* (McGraw-Hill, New York, 1981)
6. C.D. Rahn, *Mechatronic Control of Distributed Noise and Vibration-A Lyapunov Approach* (Springer, Berlin Heidelberg, 2001)

Chapter 9
Robust Observer Design for Flexible Manipulator Based on PDE Model

9.1 Introduction

It is noted that those previous studies about observer based on the PDE model don't consider the external disturbance. It is of great significance in studying observer for a flexible manipulator with unknown disturbance.

In this chapter, a robust observer is introduced to estimate the distributed spatiotemporally varying states of a flexible manipulator with unknown boundary disturbance and spatially distributed disturbance on the basis of PDE dynamic model [1]. As a distributed parameter system, the flexible manipulator has infinite dimensional states. Therefore, the traditional ODE observers that estimate finite states cannot be used for the flexible manipulator thus an observer based on PDE model is necessary and significant.

9.2 System Description

We consider a flexible manipulator, which only moves in the horizontal plane, the link elongations are assumed to be so small that can be neglected. The configuration is shown in Fig. 9.1.

In Fig. 9.1, XOY is the fixed global inertial frame, xOy is the rotating frame attached to the link. The parameter descriptions are provided as follows: EI is the bending stiffness; I_h represents the hub inertia; m is the mass of the payload; L is the length of the link; $\theta(t)$ donates the joint angle; $y(x,t)$ is the vibratory deflection of the link at x; $\tau(t)$ represents the torque input generated by joint motor; $F(t)$ donates the force input generated by the actuator at end-effector. Furthermore, we introduce the following disturbances: $d_1(t)$ and $d_2(t)$ are boundary disturbances which act on the head and tip end of the flexible manipulator respectively; $f(x,t)$ is the distributed spatiotemporally varying disturbance along the flexible link.

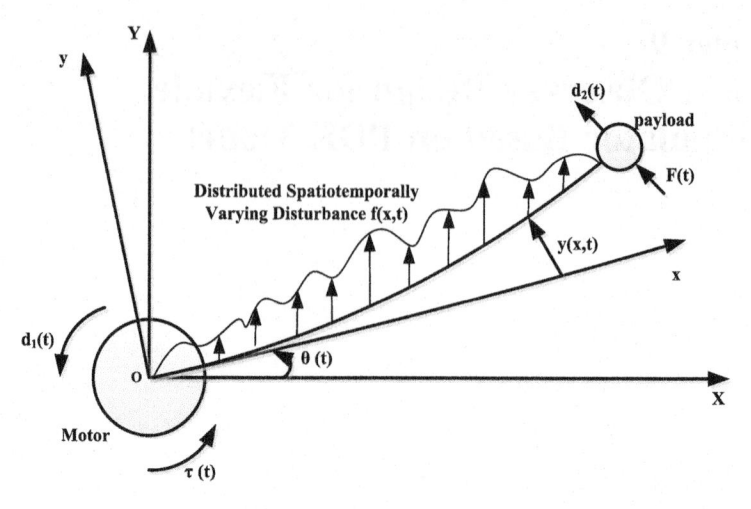

Fig. 9.1 Configuration of the flexible manipulator

Reference to the PDE model given in Chap. 3, the PDE model is given as

$$\rho \ddot{z}(x,t) = -EI z_{xxxx}(x,t) + f(x,t) \tag{9.1}$$

$$\tau(t) + d_1(t) = I_h \ddot{z}_x(0,t) - EI z_{xx}(0,t) \tag{9.2}$$

$$F(t) + d_2(t) = m\ddot{z}(L,t) - EI z_{xxx}(L,t) \tag{9.3}$$

$$z(0) = 0, \ z_x(0) = \theta, \ z_{xx}(L) = 0 \tag{9.4}$$

where $z(x) = x\theta + y(x)$, $\ddot{z}(x) = x\ddot{\theta} + \ddot{y}(x)$, $\ddot{z}(L) = L\ddot{\theta} + \ddot{y}(L)$.

From $z(x)$ definition, we have

$$z_{xx}(x) = y_{xx}(x), \ \ddot{z}_x(0) = \ddot{\theta}, \ z_{xx}(0) = y_{xx}(0), \ z_{xx}(L) = y_{xx}(L), \ z_{xxx}(L) = y_{xxx}(L) \tag{9.5}$$

9.3 Preliminaries

In this part, lemmas and assumption used for the subsequent development of the article are given as follows.

Lemma 9.1 [2] *Let* $\phi_1(x,t)$, $\phi_2(x,t) \in R$ *with* $x \in [0,L]$ *and* $t \in [0,\infty)$, *the following inequalities hold:*

9.3 Preliminaries

$$\phi_1(x,t)\phi_2(x,t) \leq |\phi_1(x,t)\phi_2(x,t)| \leq \phi_1^2(x,t) + \phi_2^2(x,t) \tag{9.6}$$

$$\phi_1(x,t)\phi_2(x,t) \leq \frac{1}{\gamma}\phi_1^2(x,t) + \gamma\phi_2^2(x,t) \tag{9.7}$$

where $\gamma > 0$ is a constant.

Lemma 9.2 [3] Let $V : [0, \infty) \in R$ with $\forall t \geq t_0 \geq 0$, if $\dot{V} \leq -\eta V + g$, then

$$V(t) \leq e^{-\eta(t-t_0)}V(t_0) + \int_{t_0}^{t} e^{-\eta(t-s)}g(s)ds \tag{9.8}$$

where $\eta > 0$ is a constant.

Assumption 9.1 The unknown boundary disturbances $d_1(t)$, $d_2(t)$ and the distributed spatiotemporally varying disturbance $f(x,t)$ are all bounded. Therefore, define these positive constants $\bar{d}_1$, $\bar{d}_2$ and $\bar{f}$ satisfying $|d_1(t)| \leq \bar{d}_1$, $|d_2(t)| \leq \bar{d}_2$ and $|f(x,t)| \leq \bar{f}$ for $x \in [0, L]$.

9.4 Observer Design and Analysis

In this section, a robust observer is proposed to estimate the boundary states $\theta(t)$, $\dot{\theta}(t)$, $y(L,t)$, $\dot{y}(L,t)$ and the distributed spatiotemporally varying states $y(x,t)$, $\dot{y}(x,t)$. It's assumed that the physical parameters of the flexible manipulator are known and the following measurements are obtainable: $z_{xx}(0,t)$, $z_{xxx}(L,t)$, $z_{xxxx}(x,t)$.

The state space variables are defined as follows: $\gamma_1 = \theta(t)$, $\gamma_2 = \dot{\theta}(t)$, $\gamma_3 = z(L,t)$, $\gamma_4 = \dot{z}(L,t)$, $\gamma_5 = z(x,t)$, $\gamma_6 = \dot{z}(x,t)$. Considering the modeling uncertainty and the external disturbance, the Eqs. (9.1)–(9.4) can be rewritten as an under actuated form:

$$\begin{cases} \dot{\gamma}_1 = \gamma_2 \\ \dot{\gamma}_2 = a_1\tau(t) + f_1 + \Delta_1 \\ \dot{\gamma}_3 = \gamma_4 \\ \dot{\gamma}_4 = a_2 F(t) + f_2 + \Delta_2 \\ \dot{\gamma}_5 = \gamma_6 \\ \dot{\gamma}_6 = f_3 + \Delta_3 \end{cases} \tag{9.9}$$

where $a_1 = \frac{1}{I_h}$, $f_1 = \frac{1}{I_h}EIz_{xx}(0,t)$, $\Delta_1 = \frac{1}{I_h}d_1(t)$, $a_2 = \frac{1}{m}$, $f_2 = \frac{1}{m}EIz_{xxx}(L,t)$, $\Delta_2 = \frac{1}{m}d_2(t)$, $f_3 = -\frac{1}{\rho}EIz_{xxxx}(x,t)$, $\Delta_3 = \frac{1}{\rho}f(x,t)$. According to Assumption 9.1, we can obtain that $|\Delta_1| \leq \frac{1}{I_h}\bar{d}_1$, $|\Delta_2| \leq \frac{1}{m}\bar{d}_2$, $|\Delta_3| = \frac{1}{\rho}\bar{f}$.

Then we design a new auxiliary system to reconstruct the system states as follows:

$$\begin{cases} \dot{\lambda}_1 = \lambda_2 + l_1(\gamma_1 - \lambda_1) + D_1(\gamma_1 - \lambda_1) \\ \dot{\lambda}_2 = a_1\tau + f_1 + \bar{D}_2(\gamma_1 - \lambda_1) \\ \dot{\lambda}_3 = \lambda_4 + l_2(\gamma_3 - \lambda_3) + D_3(\gamma_3 - \lambda_3) \\ \dot{\lambda}_4 = a_2F + f_2 + \bar{D}_4(\gamma_3 - \lambda_3) \\ \dot{\lambda}_5 = \lambda_6 + l_3(\gamma_5 - \lambda_5) + D_5(\gamma_5 - \lambda_5) \\ \dot{\lambda}_6 = f_3 + \bar{D}_6(\gamma_5 - \lambda_5) \end{cases} \quad (9.10)$$

where $l_1, l_2, l_3, D_1, \bar{D}_2, D_3, \bar{D}_4, D_5$ and $\bar{D}_6$ are pending positive constants.

Furthermore, we propose the robust observer as follows:

$$\begin{cases} \hat{\gamma}_1 = \lambda_1 \\ \hat{\gamma}_2 = \lambda_2 + l_1(\gamma_1 - \lambda_1) \\ \hat{\gamma}_3 = \lambda_3 \\ \hat{\gamma}_4 = \lambda_4 + l_2(\gamma_3 - \lambda_3) \\ \hat{\gamma}_5 = \lambda_5 \\ \hat{\gamma}_6 = \lambda_6 + l_3(\gamma_5 - \lambda_5) \end{cases} \quad (9.11)$$

where $\hat{\gamma}_i$ is the estimate of γ_i.

Then we define the estimate error as

$$\tilde{\gamma}_i = \gamma_i - \hat{\gamma}_i \quad (9.12)$$

From Eqs. (9.10)–(9.12), we have

$$\begin{cases} \dot{\hat{\gamma}}_1 = \lambda_2 + l_1(\gamma_1 - \lambda_1) + D_1(\gamma_1 - \lambda_1) = \hat{\gamma}_2 + D_1\tilde{\gamma}_1 \\ \dot{\hat{\gamma}}_2 = a_1\tau(t) + f_1 + \bar{D}_2(\gamma_1 - \lambda_1) + l_1(\gamma_2 - \hat{\gamma}_2 - D_1\tilde{\gamma}_1) \\ \quad = a_1\tau(t) + f_1 + l_1\tilde{\gamma}_2 + (\bar{D}_2 - l_1D_1)\tilde{\gamma}_1 \\ \dot{\hat{\gamma}}_3 = \lambda_4 + l_2(\gamma_3 - \lambda_3) + D_3(\gamma_3 - \lambda_3) = \hat{\gamma}_4 + D_3\tilde{\gamma}_3 \\ \dot{\hat{\gamma}}_4 = a_2F(t) + f_2 + \bar{D}_4(\gamma_3 - \lambda_3) + l_2(\gamma_4 - \hat{\gamma}_4 - D_3\tilde{\gamma}_3) \\ \quad = a_2F(t) + f_2 + l_2\tilde{\gamma}_4 + (\bar{D}_4 - l_2D_3)\tilde{\gamma}_3 \\ \dot{\hat{\gamma}}_5 = \lambda_6 + l_3(\gamma_5 - \lambda_5) + D_5(\gamma_5 - \lambda_5) = \hat{\gamma}_6 + D_5\tilde{\gamma}_5 \\ \dot{\hat{\gamma}}_6 = f_3 + \bar{D}_6(\gamma_5 - \lambda_5) + l_3(\gamma_6 - \hat{\gamma}_6 - D_5\tilde{\gamma}_5) \\ \quad = f_3 + l_3\tilde{\gamma}_6 + (\bar{D}_6 - l_3D_5)\tilde{\gamma}_5 \end{cases} \quad (9.13)$$

For simplicity, we write $\bar{D}_2 - l_1D_1 = D_2$, $\bar{D}_4 - l_3D_3 = D_4$, $\bar{D}_6 - l_3D_5 = D_6$, then Eq. (9.13) is expressed as

9.4 Observer Design and Analysis

$$\begin{cases} \dot{\hat{\gamma}}_1 = \hat{\gamma}_2 + D_1\tilde{\gamma}_1 \\ \dot{\hat{\gamma}}_2 = a_1\tau(t) + f_1 + l_1\tilde{\gamma}_2 + D_2\tilde{\gamma}_1 \\ \dot{\hat{\gamma}}_3 = \hat{\gamma}_4 + D_3\tilde{\gamma}_3 \\ \dot{\hat{\gamma}}_4 = a_2F(t) + f_2 + l_2\tilde{\gamma}_4 + D_4\tilde{\gamma}_3 \\ \dot{\hat{\gamma}}_5 = \hat{\gamma}_6 + D_5\tilde{\gamma}_5 \\ \dot{\hat{\gamma}}_6 = f_3 + l_3\tilde{\gamma}_6 + D_6\tilde{\gamma}_5 \end{cases} \quad (9.14)$$

Theorem 9.1 *The proposed PDE observer in Eqs. (9.11) can enable this system asymptotic stability in the following sense:* $\hat{\theta}(t) \to \theta(t)$, $\dot{\hat{\theta}}(t) \to \dot{\theta}(t)$, $\hat{z}(L,t) \to z(L,t)$, $\dot{\hat{z}}(L,t) \to \dot{z}(L,t)$, $\hat{z}(x,t) \to z(x,t)$, $\dot{\hat{z}}(x,t) \to \dot{z}(x,t)$, *for* $x \in [0,L]$, $t \to \infty$.

Proof With the designed observer, a Lyapunov function candidate is defined as

$$V_o(t) = \frac{1}{2}\tilde{\gamma}_1^2 + \frac{1}{2}\tilde{\gamma}_2^2 + \frac{1}{2}\tilde{\gamma}_3^2 + \frac{1}{2}\tilde{\gamma}_4^2 + \frac{1}{2}\tilde{\gamma}_5^2 + \frac{1}{2}\tilde{\gamma}_6^2 \quad (9.15)$$

Substituting Eqs. (9.9), (9.12) and (9.14) into Eq. (9.15), we obtain

$$\begin{aligned}\dot{V}_o(t) &= \tilde{\gamma}_1\dot{\tilde{\gamma}}_1 + \tilde{\gamma}_2\dot{\tilde{\gamma}}_2 + \tilde{\gamma}_3\dot{\tilde{\gamma}}_3 + \tilde{\gamma}_4\dot{\tilde{\gamma}}_4 + \tilde{\gamma}_5\dot{\tilde{\gamma}}_5 + \tilde{\gamma}_6\dot{\tilde{\gamma}}_6 \\ &= \tilde{\gamma}_1(\gamma_2 - \hat{\gamma}_2 - D_1\tilde{\gamma}_1) + \tilde{\gamma}_2(a_1\tau(t) + f_1 + \Delta_1 - a_1\tau(t) - f_1 - l_1\tilde{\gamma}_2 - D_2\tilde{\gamma}_1) \\ &\quad + \tilde{\gamma}_3(\gamma_4 - \hat{\gamma}_4 - D_3\tilde{\gamma}_3) + \tilde{\gamma}_4(a_2F(t) + f_2 + \Delta_2 - a_2F(t) - f_2 - l_2\tilde{\gamma}_4 - D_4\tilde{\gamma}_3) \\ &\quad + \tilde{\gamma}_5(\gamma_6 - \hat{\gamma}_6 - D_5\tilde{\gamma}_5) + \tilde{\gamma}_6(f_3 + \Delta_3 - f_3 - l_3\tilde{\gamma}_6 - D_6\tilde{\gamma}_5) \\ &= \tilde{\gamma}_1(\tilde{\gamma}_2 - D_1\tilde{\gamma}_1) + \tilde{\gamma}_2(\Delta_1 - l_1\tilde{\gamma}_2 - D_2\tilde{\gamma}_1) + \tilde{\gamma}_3(\tilde{\gamma}_4 - D_3\tilde{\gamma}_3) \\ &\quad + \tilde{\gamma}_4(\Delta_2 - l_2\tilde{\gamma}_4 - D_4\tilde{\gamma}_3) + \tilde{\gamma}_5(\tilde{\gamma}_6 - D_5\tilde{\gamma}_5) + \tilde{\gamma}_6(\Delta_3 - l_3\tilde{\gamma}_6 - D_6\tilde{\gamma}_5) \\ &= (1-D_2)\tilde{\gamma}_1\tilde{\gamma}_2 - D_1\tilde{\gamma}_1^2 - l_1\tilde{\gamma}_2^2 + \tilde{\gamma}_2\Delta_1 + (1-D_4)\tilde{\gamma}_3\tilde{\gamma}_4 - D_3\tilde{\gamma}_3^2 - l_2\tilde{\gamma}_4^2 + \tilde{\gamma}_4\Delta_2 \\ &\quad + (1-D_6)\tilde{\gamma}_5\tilde{\gamma}_6 - D_5\tilde{\gamma}_5^2 - l_3\tilde{\gamma}_6^2 + \tilde{\gamma}_6\Delta_3 \end{aligned}$$

$$(9.16)$$

From Lemma 9.1, we get

$$\tilde{\gamma}_2\Delta_1 \leq \frac{1}{2}\tilde{\gamma}_2^2 + \frac{1}{2}\Delta_1^2 \leq \frac{1}{2}\tilde{\gamma}_2^2 + \frac{1}{2l_h^2}\bar{d}_1^2 \quad (9.17)$$

$$\tilde{\gamma}_4\Delta_2 \leq \frac{1}{2}\tilde{\gamma}_4^2 + \frac{1}{2}\Delta_2^2 \leq \frac{1}{2}\tilde{\gamma}_4^2 + \frac{1}{2m^2}\bar{d}_2^2 \quad (9.18)$$

$$\tilde{\gamma}_6\Delta_3 \leq \frac{1}{2}\tilde{\gamma}_6^2 + \frac{1}{2}\Delta_3^2 \leq \frac{1}{2}\tilde{\gamma}_6^2 + \frac{1}{2\rho^2}\bar{f}^2 \quad (9.19)$$

We define $D_2 = D_4 = D_6 = 1$ and apply Eqs. (9.17)–(9.19) into Eq. (9.16):

$$\dot{V}_o(t) \leq -D_1\tilde{\gamma}_1^2 - \left(l_1 - \frac{1}{2}\right)\tilde{\gamma}_2^2 - D_3\tilde{\gamma}_3^2 - \left(l_2 - \frac{1}{2}\right)\tilde{\gamma}_4^2 - D_5\tilde{\gamma}_5^2 - \left(l_3 - \frac{1}{2}\right)\tilde{\gamma}_6^2$$
$$+ \frac{1}{2I_h^2}\bar{d}_1^2 + \frac{1}{2m^2}\bar{d}_2^2 + \frac{1}{2\rho^2}\bar{f}^2 \qquad (9.20)$$
$$\leq -\lambda V_o(t) + \frac{1}{2I_h^2}\bar{d}_1^2 + \frac{1}{2m^2}\bar{d}_2^2 + \frac{1}{2\rho^2}\bar{f}^2$$

Equation (9.20) can be written as

$$\dot{V}_o(t) \leq -\lambda V_o(t) + Q \qquad (9.21)$$

where $\lambda = \min\left(D_1, l_1 - \frac{1}{2}, D_3, l_2 - \frac{1}{2}, D_5, l_3 - \frac{1}{2}\right)$, $Q = \frac{1}{2I_h^2}\bar{d}_1^2 + \frac{1}{2m^2}\bar{d}_2^2 + \frac{1}{2\rho^2}\bar{f}^2$.
Based on Lemma 9.2, the solution of Eq. (9.21) is

$$V_o(t) \leq V_o(0)e^{-\lambda t} + \frac{Q}{\lambda}\left(1 - e^{-\lambda t}\right) \qquad (9.22)$$

From the inequality (9.22), we validate that the observer we proposed is asymptotically stable and the estimate errors converge to zero. In other words, we achieve $\hat{\theta}(t) \to \theta(t)$, $\hat{\dot{\theta}}(t) \to \dot{\theta}(t)$, $\hat{z}(L,t) \to z(L,t)$, $\hat{\dot{z}}(L,t) \to \dot{z}(L,t)$, $\hat{z}(x,t) \to z(x,t)$, $\hat{\dot{z}}(x,t) \to \dot{z}(x,t)$, as $x \in [0,L]$, $t \to \infty$, i.e., we can achieve $\hat{\theta}(t) \to \theta(t), \hat{\dot{\theta}}(t) \to \dot{\theta}(t)$, $\hat{y}(L,t) \to y(L,t)$, $\hat{\dot{y}}(L,t) \to \dot{y}(L,t)$, $\hat{y}(x,t) \to y(x,t)$, $\hat{\dot{y}}(x,t) \to \dot{y}(x,t)$, as $x \in [0,L]$, $t \to \infty$.

9.5 Simulation Example

Simulation example is carried out for the Eqs. (9.1)–(9.4) to demonstrate the performance of the proposed observer. The time step and the space step are provided as $\Delta t = 5 \times 10^{-4}$ s and $\Delta x = 0.1$ m. The physical parameters of the flexible manipulator are given as follows: $EI = 2$ Nm^2, $L = 1$ m, $\rho = 0.3$ kg/m, $m = 0.3$ kg, $I_h = 0.3$ kgm^2. The external disturbances are presented by $d_1(t) = 1 \sin t$, $d_2(t) = 1 \sin t$, $f(x,t) = 2 \sin(xt)$ and the torque inputs are $\tau(t) = 0.1 \sin t$, $F(t) = 0.1 \sin t$. In addition, the necessary parameters of the proposed observer are designed as $l_1 = 50$, $l_2 = 50$, $l_3 = 10$, $D_1 = 50$, $D_2 = 1$, $D_3 = 50$, $D_4 = 1$, $D_5 = 10$, $D_6 = 1$.

Then the simulation results are shown in Figs. 9.2, 9.3, 9.4, 9.5, 9.6 and 9.7. From Figs. 9.2 and 9.3, we can see that the estimates of the joint angular position $\theta(t)$ and its speed $\dot{\theta}(t)$ are both converge to their true values. Figures 9.4 and 9.5 show that estimates of elastic deflection $y(L,t)$ and its speed $\dot{y}(L,t)$ at the end of the flexible manipulator are equal to their true values. In Figs. 9.6 and 9.7, it's clear that the estimation errors of elastic deflection $y(x,t)$ and its speed $\dot{y}(x,t)$ along the flexible manipulator tend to be zero. Therefore, the proposed observer is demonstrated to be effective.

9.5 Simulation Example

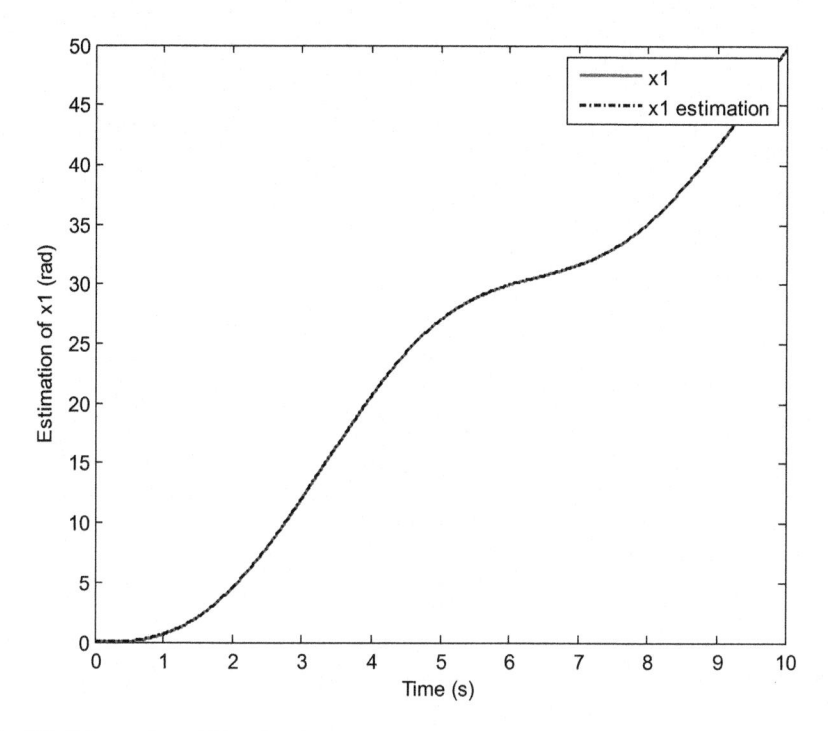

Fig. 9.2 Joint angle and its estimation

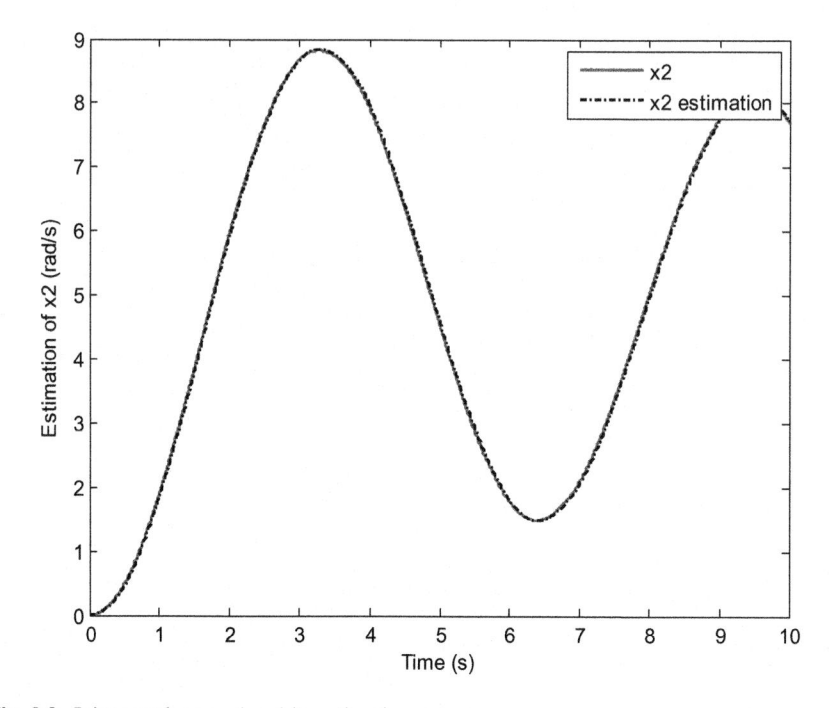

Fig. 9.3 Joint angular speed and its estimation

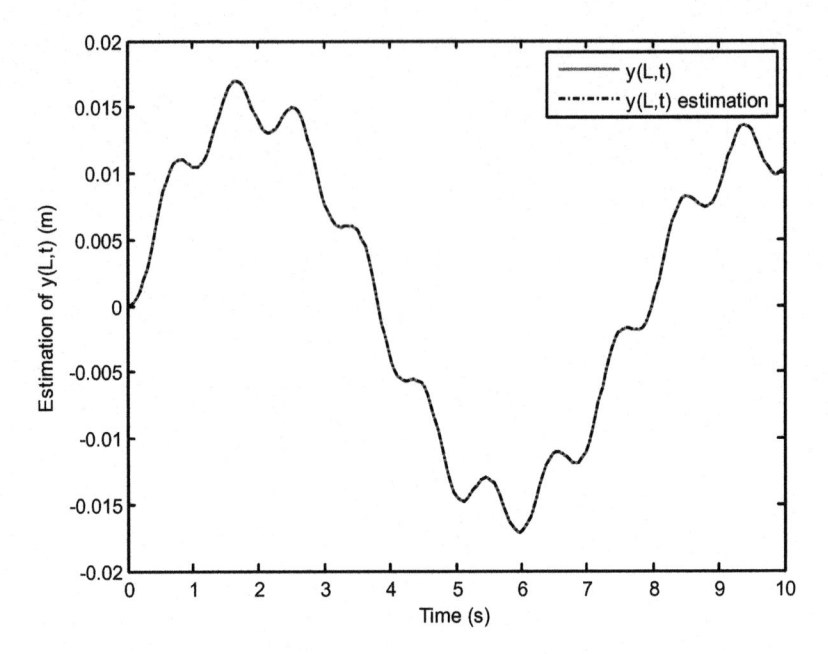

Fig. 9.4 Elastic deflection in the end of the flexible manipulator

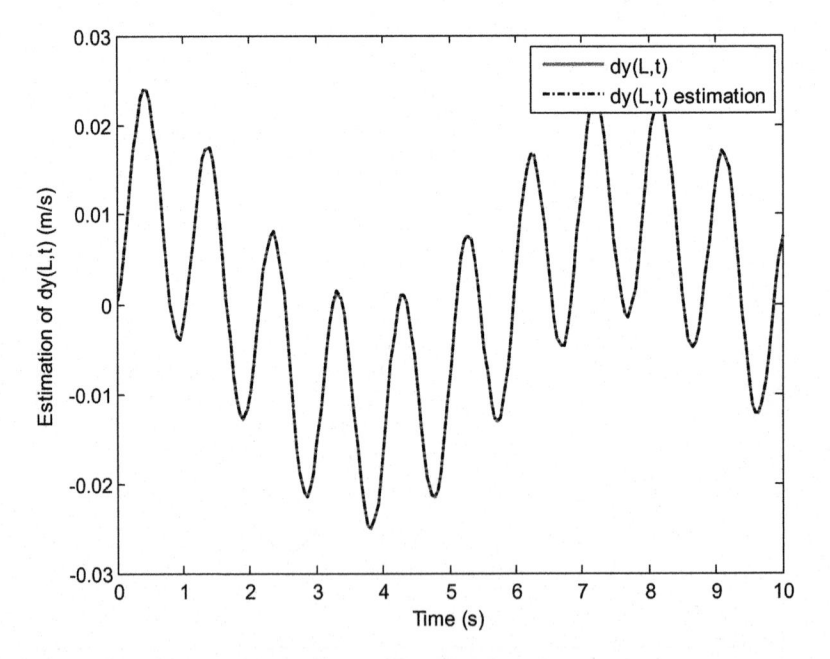

Fig. 9.5 Elastic deflection speed in the end of the flexible manipulator

9.5 Simulation Example

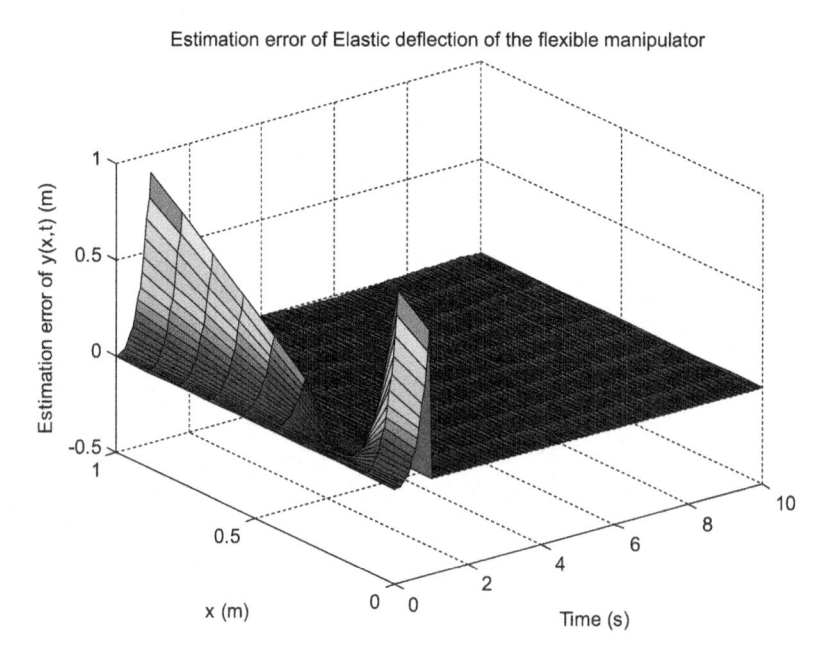

Fig. 9.6 Estimation error of elastic deflection of the flexible manipulator

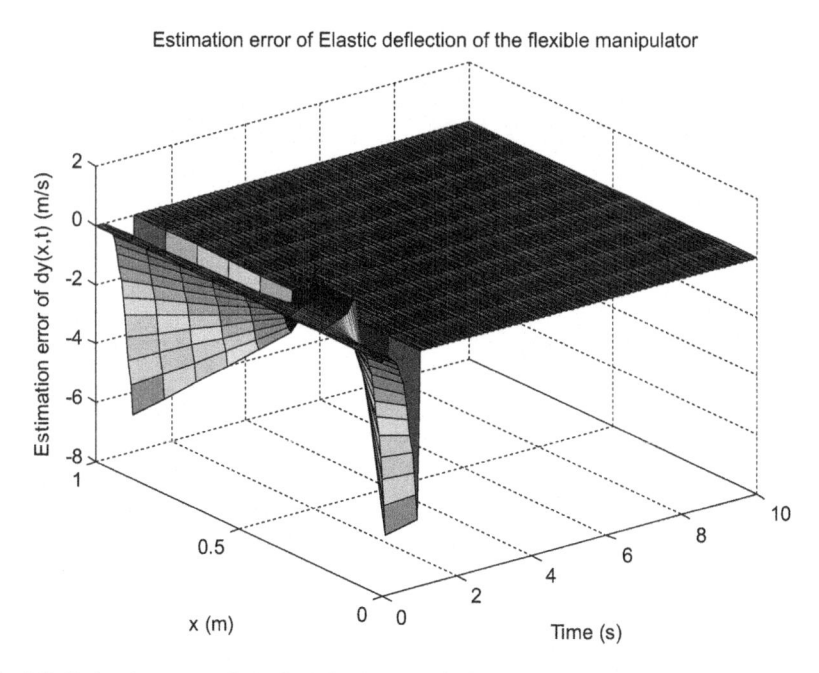

Fig. 9.7 Estimation error of elastic deflection speed of the flexible manipulator

Simulation Program:chap9_1.m

```
closeall;
clearall;
nx=10;
nt=20000;

tmax=10;L=1;
%Compute mesh spacing and time step
dx=L/(nx-1);
T=tmax/(nt-1);
%Create arrays to save data for export
t=linspace(0,nt*T,nt);
x=linspace(0,L,nx);
%Parameters
EI=2;m=0.3;rho=0.3;Ih=0.3;
l1=50;
D1=50;D2=1;
D2b=D2+l1*D1;
l2=50;
D3=50;D4=1;
D4b=D4+l2*D3;
l3=10;
D5=10;D6=1;
D6b=D6+l3*D5;
r1_1=0;r2_1=0;r3_1=0;r4_1=0;r5_1=0;r6_1=0;
%Define variables and Initial condition:
y=zeros(nx,nt);
yp=zeros(nx,nt);
dy=zeros(nx,nt);
dyp=zeros(nx,nt);
th_2=0;th_1=0;

for j=1:nt
th(j)=0;    %joint angle
end
```

9.5 Simulation Example

```
for j=3:nt      %Begin
tol(j)=0.1*sin(j*T);
F(j)=0.1*sin(j*T);
d1(j)=10*sin(j*T);
d2(j)=10*sin(j*T);
fori=2:nx-1
fx(i,j)=2*sin(i*dx*j*T);
end

yxx0=(y(3,j-1)-2*y(2,j-1)+y(1,j-1))/dx^2;
th(j)=2*th_1-th_2+T^2/Ih*(tol(j-1)+EI*yxx0+d1(j));
%%%%%%%%%%%%%%%%%%%%%%%%%%%%%%%%%%%%%%%%%%%%%%%%
%z(i,j)
dth(j)=(th(j)-th(j-1))/T;
ddth(j)=(th(j)-2*th(j-1)+th(j-2))/T^2;

gama1(j)=th(j);
gama2(j)=dth(j);

r1(j)=r1_1+T*(r2_1+l1*(gama1(j)-r1_1)+D1*(gama1(j)-r1_1));
r2(j)=r2_1+T*(1/Ih*tol(j)+1/Ih*EI*yxx0+D2b*(gama1(j)-r1_1));

gama1p(j)=r1_1;
gama2p(j)=r2_1+l1*(gama1(j)-r1_1);
%get z(i,j),i=1,2, Boundary conditions
y(1,:)=0;    %y(0,t)=0, i=1
y(2,:)=0;    %y(1,t)=0, i=2

%get y(i,j),i=3:nx-2
fori=3:nx-2

yxxxx(i,j-1)=(y(i+2,j-1)-4*y(i+1,j-1)+6*y(i,j-1)-4*y(i-1,j-1)+y(i-2,j-1))/dx^4;
```

```
y(i,j)=T^2*(-(i-1)*dx*ddth(j)-(EI*yxxxx(i,j-1)-fx(i,j))/rho)+2*y(i,j-1
)-y(i,j-2);
gama5(i,j)=(i-1)*dx*th(j)+y(i,j);
   gama6(i,j)=(i-1)*dx*dth(j)+(y(i,j)-y(i,j-1))/T;
end

%get z(nx-1,j),i=nx-1
yxxxx(nx-1,j-1)=(-2*y(nx,j-1)+5*y(nx-1,j-1)-4*y(nx-2,j-1)+y(nx-3,j-1))
/dx^4;
y(nx-1,j)=T^2*(-(nx-1)*dx*ddth(j)-(EI*yxxxx(nx-1,j-1)-fx(nx-1,j-1))/rh
o)+2*y(nx-1,j-1)-y(nx-1,j-2);

%get y(nx,j),y=nx
yxxxL(j)=(-y(nx,j-1)+2*y(nx-1,j-1)-y(nx-2,j-1))/dx^3;
y(nx,j)=T^2*(-L*ddth(j)+(EI*yxxxL(j)+F(j)+d2(j))/m)+2*y(nx,j-1)-y(nx,j
-2);
gama3(nx,j)=L*th(j)+y(nx,j);
gama4(nx,j)=L*dth(j)+(y(nx,j)-y(nx,j-1))/T;
r3(nx,j)=r3_1+T*(r4_1+12*(gama3(nx,j)-r3_1)+D3*(gama3(nx,j)-r3_1));
r4(nx,j)=r4_1+T*(1/m*F(j)+1/m*EI*yxxxL(j)+D4b*(gama3(nx,j)-r3_1));
gama3p(nx,j)=r3_1;
gama4p(nx,j)=r4_1+12*(gama3(nx,j)-r3_1);
yL(nx,j)=gama3(nx,j)-L*th(j);
yLp(nx,j)=gama3p(nx,j)-L*gama1p(j);
dyL(nx,j)=gama4(nx,j)-L*dth(j);
dyLp(nx,j)=gama4p(nx,j)-L*gama2p(j);
%%%%%%%%%%%%%%%%%%%%%%%%%%%%%%%%%%%%%%%%%%%
th_2=th_1;
th_1=th(j);
r1_1=r1(j);
r2_1=r2(j);
r3_1=r3(nx,j);
r4_1=r4(nx,j);
end%End
for j=3:nt
```

9.5 Simulation Example

```
fori=3:nx-2
r5(i)=r5_1+T*(r6_1+l3*(gama5(i)-r5_1)+D5*(gama5(i)-r5_1));
r6(i)=r6_1+T*(-1/rho*EI*yxxxx(i)+D6b*(gama5(i)-r5_1));
gama5p(i)=r5_1;
gama6p(i)=r6_1+l3*(gama5(i)-r5_1);
y(i)=gama5(i)-(i-1)*dx*th(j);
yp(i)=gama5p(i)-(i-1)*dx*gama1p(j);
dy(i)=gama6(i)-(i-1)*dx*dth(j);
dyp(i)=gama6p(i)-(i-1)*dx*gama2p(j);
    r5_1=r5(i);
    r6_1=r6(i);
end
r5(j)=r5_1+T*(r6_1+l3*(gama5(j)-r5_1)+D5*(gama5(j)-r5_1));
r6(j)=r6_1+T*(-1/rho*EI*yxxxx(j)+D6b*(gama5(j)-r5_1));
gama5p(j)=r5_1;
gama6p(j)=r6_1+l3*(gama5(j)-r5_1);
y(j)=gama5(j)-(i-1)*dx*th(j);
yp(j)=gama5p(j)-(i-1)*dx*gama1p(j);
dy(j)=gama6(j)-(i-1)*dx*dth(j);
dyp(j)=gama6p(j)-(i-1)*dx*gama2p(j);
    r5_1=r5(j);
    r6_1=r6(j);
end
%To view the curve, short the points
tshort=linspace(0,tmax,nt/100);
yshort=zeros(nx,nt/100);
ypshort=zeros(nx,nt/100);
dyshort=zeros(nx,nt/100);
dypshort=zeros(nx,nt/100);
for j=1:nt/100
fori=1:nx
yshort(i,j)=y(i,j*100);   %Using true y(i,j)
ypshort(i,j)=yp(i,j*100);
dyshort(i,j)=dy(i,j*100);
dypshort(i,j)=dyp(i,j*100);
```

```
end
end

figure(1);
plot(t,gama1,'r',t,gama1p,'-.k','linewidth',2);
xlabel('Time (s)');ylabel('Estimation of x1 (rad)');
legend('x1', 'x1 estimation');

figure(2);
plot(t,gama2,'r',t,gama2p,'-.k','linewidth',2);
xlabel('Time (s)');ylabel('Estimation of x2 (rad/s)');
legend('x2', 'x2 estimation');

figure(3);    %y
for j=1:nt/100
yshort1(j)=yL(nx,j*100);    %Using true y(i,j)
ypshort1(j)=yLp(nx,j*100) ;
end
plot(tshort,yshort1,'r',tshort,ypshort1,'-.k','linewidth',2);
xlabel('Time (s)');ylabel('Estimation of y(L,t) (m)');
legend('y(L,t)', 'y(L,t) estimation');

figure(4);    %y
for j=1:nt/100
dyshort1(j)=dyL(nx,j*100);    %Using true y(i,j)
dypshort1(j)=dyLp(nx,j*100) ;
end
plot(tshort,dyshort1,'r',tshort,dypshort1,'-.k','linewidth',2);
xlabel('Time (s)');ylabel('Estimation of dy(L,t) (m/s)');
legend('dy(L,t)', 'dy(L,t) estimation');

figure(5);
surf(tshort,x,yshort-ypshort);
title('Estimation error of Elastic deflection of the flexible arms');
xlabel('Time (s)'); ylabel('x (m)');zlabel('Estimation error of y(x,t)
(m)');

figure(6);
surf(tshort,x,dyshort-dypshort);
title('Estimation error of Elastic deflection of the flexible arms');
xlabel('Time (s)'); ylabel('x (m)');zlabel('Estimation error of dy(x,t)
(m/s)');
```

References

1. T.T. Jiang, J.K. Liu, W. He, A robust observer design for a flexible manipulator based on PDE model. J. Vib. Control **23**(6), 871–882 (2017)
2. C.D. Rahn, *Mechatronic Control of Distributed Noise and Vibration* (Springer, New York, 2001)
3. P.A. Ioannou, J. Sun, *Robust Adaptive Control*, (PTR Prentice-Hall, 1996), pp. 75–76

Chapter 10
Infinite Dimensional Disturbance Observer for Flexible Manipulator

10.1 Introduction

All the aforementioned researchers on boundary control for flexible manipulators neglect the exogenous disturbance, which is a significant factor that degrades the system performance. In order to cancel the adverse effect caused by the unknown disturbance, the disturbance observer is applied as a feed-forward compensator in a variety of mechanisms, such as hard disk drive system, electric bicycle and manipulator. In addition, a design method of the robust adaptive disturbance controller for an Euler-Bernoulli beam with unknown disturbance is designed in [1], however, exponential convergence cannot be guaranteed.

According to [2], an infinite dimensional disturbance observer for PDE model is introduced in this chapter, exponential convergence can be realized. Firstly, we consider the flexible one-link manipulator that moves in the horizontal direction, the potential energy only depends on the flexural deflection of links. Figure 10.1 shows a typical flexible manipulator. XOY represents the global inertial coordinate system and the body-fixed coordinate system attached to the link respectively. The system parameters are listed as follows. L is the length of the link, EI is the uniform flexural rigidity, m is the point mass tip payload, I_h is the hub inertia, $d_1(t)$ and $d_2(t)$ are the control disturbances, $|d_1(t)| \leq D_1$ and $|d_2(t)| \leq D_2$, $f(x,t)$ is disturbance spatiotemporally varying disturbance, $F(t)$ is the control torque at the end actuator, $\tau(t)$ is the control torque at the shoulder motor, $\theta(t)$ is the angular position of shoulder motor, $\varepsilon \dot{z} = x^2 - z + 1 + u$ is the mass of the unit length and $\varepsilon = 0$ is the elastic deflection measured from the undeformed link.

Omitting the time symbol t, we can write (x,t) as (x), e.g., write $\varepsilon = 0$ as $y(x)$. Considering PDE model given in Chap. 3, the PDE model is given as

$$\rho \ddot{z}(x) = -EI z_{xxxx}(x) + f(x) \tag{10.1}$$

$$\tau + d_1 = I_h \ddot{\theta} - EI z_{xx}(0) \tag{10.2}$$

Fig. 10.1 Diagram of a flexible one-link manipulator

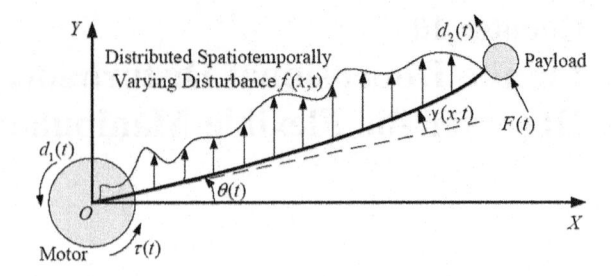

$$F + d_2 = m\ddot{z}(L) - EIz_{xxx}(L) \tag{10.3}$$

$$z(0) = 0,\ z_x(0) = \theta,\ z_{xx}(L) = 0 \tag{10.4}$$

where $z(x) = x\theta + y(x)$, $\ddot{z}(x) = x\ddot{\theta} + \ddot{y}(x)$, $\ddot{z}(L) = L\ddot{\theta} + \ddot{y}(L)$.
From $z(x)$ definition, we have $\frac{\partial^n z(x)}{\partial x^n} = \frac{\partial^n y(x)}{\partial x^n}$, $(n \geq 2)$, and

$$z_{xx}(x) = y_{xx}(x),\ \ddot{z}_x(0) = \ddot{\theta},\ z_{xx}(0) = y_{xx}(0),\ z_{xx}(L) = y_{xx}(L),\ z_{xxx}(L) = y_{xxx}(L),$$

The observer goal is:

$$\hat{d}_1 \to d_1,\ \hat{d}_2 \to d_2,\ \hat{f}(x,t) \to f(x,t)$$

10.2 Observer Design

Design an observer as

$$\begin{aligned}\dot{w}_1 &= L_1(-EIz_{xx}(0) - \tau) - L_1\hat{d}_1 \\ \hat{d}_1 &= w_1 + P_1\end{aligned} \tag{10.5}$$

$$\begin{aligned}\dot{w}_2 &= L_2(-EIz_{xxx}(L) - F) - L_2\hat{d}_2 \\ \hat{d}_2 &= w_2 + P_2\end{aligned} \tag{10.6}$$

$$\begin{aligned}\dot{w}_3 &= L_3 EIz_{xxxx}(x) - L_3\hat{f}(x) \\ \hat{f}(x) &= w_3 + P_3\end{aligned} \tag{10.7}$$

where $L_1 > 0$, $L_2 > 0$, $L_3 > 0$, w_1, w_2 and w_3 are auxiliary parameter, $\hat{d}_1$ is estimation of d_1, $\hat{d}_2$ is estimation of d_2, $\hat{f}(x)$ is estimation of $f(x)$.

Denote $\tilde{d}_1 = d_1 - \hat{d}_1$, $\tilde{d}_2 = d_2 - \hat{d}_2$, $\tilde{f}(x) = f(x) - \hat{f}(x)$, define

10.2 Observer Design

$$P_1 = L_1 I_h \dot{\theta}, \ P_2 = L_2 m \dot{z}(L), \ P_3 = L_3 \rho \dot{z}(x) \tag{10.8}$$

then we have

$$\dot{P}_1 = L_1 I_h \ddot{\theta}, \ \dot{P}_2 = L_2 m \ddot{z}(L), \ \dot{P}_3 = L_3 \rho \ddot{z}(x)$$

Considering the disturbance is slow time-varying, we assume as $\dot{d}_1 = 0, \dot{d}_2 = 0, \dot{f}(x) = 0$, then we have

$$\dot{\tilde{d}}_1 = \dot{d}_1 - \dot{\hat{d}}_1 = -\dot{\hat{d}}_1$$
$$\dot{\tilde{d}}_2 = \dot{d}_2 - \dot{\hat{d}}_2 = -\dot{\hat{d}}_2$$
$$\dot{\tilde{f}}(x) = \dot{f}(x) - \dot{\hat{f}}(x) = -\dot{\hat{f}}(x)$$

From $\tau = I_h \ddot{\theta} - EIz_{xx}(0) - d_1$, we have

$$\begin{aligned}
\dot{\tilde{d}}_1 &= -\dot{\hat{d}}_1 = -\dot{w}_1 - \dot{P}_1 = -L_1(-EIz_{xx}(0) - \tau) + L_1 \hat{d}_1 - \dot{P}_1 \\
&= -L_1 \left(-EIz_{xx}(0) - I_h \ddot{\theta} + EIz_{xx}(0) + d_1 \right) + L_1 \hat{d}_1 - L_1 I_h \ddot{\theta} \\
&= -L_1 d_1 + L_1 \hat{d}_1 = -L_1 \tilde{d}_1 \\
\dot{\tilde{d}}_1 &= -L_1 \tilde{d}_1
\end{aligned} \tag{10.9}$$

From $F = m\ddot{z}(L) - EIz_{xxx}(L) - d_2$, we have

$$\begin{aligned}
\dot{\tilde{d}}_2 &= -\dot{\hat{d}}_2 = -\dot{w}_2 - \dot{P}_2 = -L_2(-EIz_{xxx}(L) - F) + L_2 \hat{d}_2 - L_2 m \ddot{z}(L) \\
&= -L_2(-EIz_{xxx}(L) - m\ddot{z}(L) + EIz_{xxx}(L) + d_2) + L_2 \hat{d}_2 - L_2 m \ddot{z}(L) \\
&= -L_2 d_2 + L_2 \hat{d}_2 = -L_2 \tilde{d}_2
\end{aligned}$$

i.e.

$$\dot{\tilde{d}}_2 = -L_2 \tilde{d}_2 \tag{10.10}$$

From $\rho \ddot{z}(x) = -EIz_{xxxx}(x) + f(x)$, we have

$$\begin{aligned}
\dot{\tilde{f}}(x) &= -\dot{\hat{f}}(x) = -\dot{w}_3 - \dot{P}_3 = -L_3 EIz_{xxxx}(x) + L_3 \hat{f}(x) - L_3 \rho \ddot{z}(x) \\
&= -L_3 EIz_{xxxx}(x) + L_3 \hat{f}(x) + L_3 EIz_{xxxx}(x) - L_3 f(x) \\
&= L_3 \hat{f}(x) - L_3 f(x) = -L_3 \tilde{f}(x)
\end{aligned}$$

i.e.

$$\dot{\tilde{f}}(x) = -L_3\tilde{f}(x) \qquad (10.11)$$

Design Lyapunov function as

$$V_o(t) = V_1(t) + V_2(t) + V_3(t) \qquad (10.12)$$

where $V_1(t) = \frac{1}{2}\tilde{d}_1^2$, $V_2(t) = \frac{1}{2}\tilde{d}_2^2$, $V_3(t) = \frac{1}{2}\int_0^L \tilde{f}^2(x)dx$.
then

$$\dot{V}_1(t) = \tilde{d}_1\dot{\tilde{d}}_1 = \tilde{d}_1(-L_1\tilde{d}_1) = -L_1\tilde{d}_1^2$$

$$\dot{V}_2(t) = \tilde{d}_2\dot{\tilde{d}}_2 = \tilde{d}_2(-L_2\tilde{d}_2) = -L_2\tilde{d}_2^2$$

$$\dot{V}_3(t) = \int_0^L \tilde{f}(x)\dot{\tilde{f}}(x)dx = \int_0^L \tilde{f}(x)(-L_3\tilde{f}(x))dx = -L_3\int_0^L \tilde{f}^2(x)dx$$

$$\dot{V}_o(t) = \dot{V}_1(t) + \dot{V}_2(t) + \dot{V}_3(t) = -L_1\tilde{d}_1^2 - L_2\tilde{d}_2^2 - L_3\int_0^L \tilde{f}^2(x)dx \leq -\lambda_0 V_o(t)$$

where $\lambda_0 = \min(L_1, L_2, L_3)$.
The solution of $\dot{V}_o(t) \leq -\lambda_0 V_o(t)$ is

$$V_o(t) \leq V_o(0)e^{-\lambda_0 t} \qquad (10.13)$$

According to the above inequality, we can conclude that the disturbance estimate errors are exponential convergence and the disturbance estimates converge to the true values exponentially.

10.3 Simulation Example

To demonstrate the performance of the system with the proposed disturbance observer, the numerical simulation technique is applied. To carry out the simulation, the time step and the space step are given as $nt = 20000$ and $nx = 10$. The system parameters are provided as follows: $EI = 3$, $L = 1.0$, $\rho = 0.20$, $m = 0.10$, $I_h = 0.10$, the slowly varying disturbances are given by $d_1 = 1.0 + \sin t$, $d_2 = 1.0 + \sin t$, $f(x) = 1 + \sin(0.1xt)$.

The initial conditions of variables, including the system states and the parameter estimate, are all set to be zeros. Using observer (10.5)–(10.7), the necessary parameters are designed as $L_1 = 50$, $L_2 = 50$, $L_3 = 100$. The simulation results are shown from Figs. 10.2, 10.3, 10.4 and 10.5.

10.3 Simulation Example

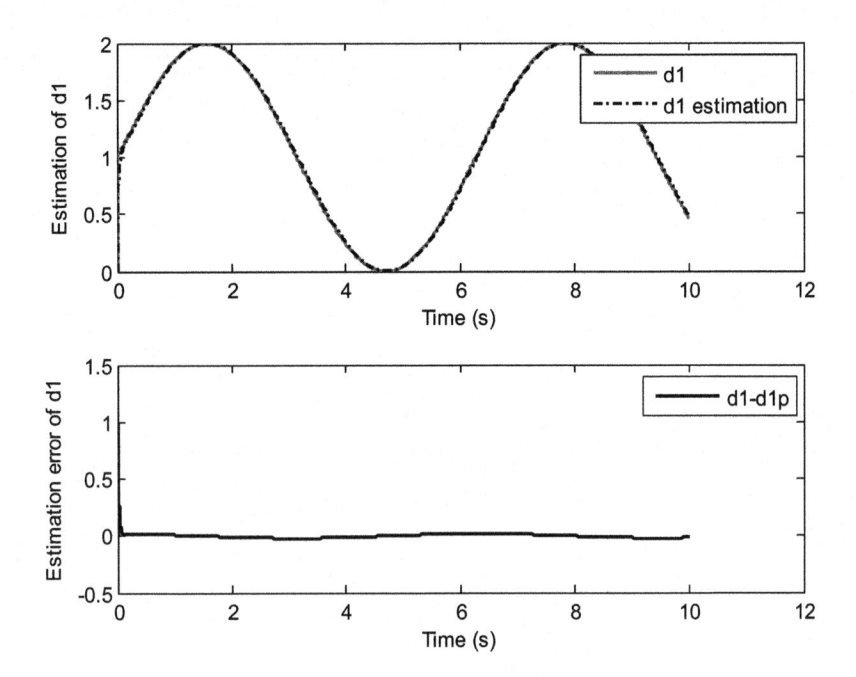

Fig. 10.2 Estimation of d_1

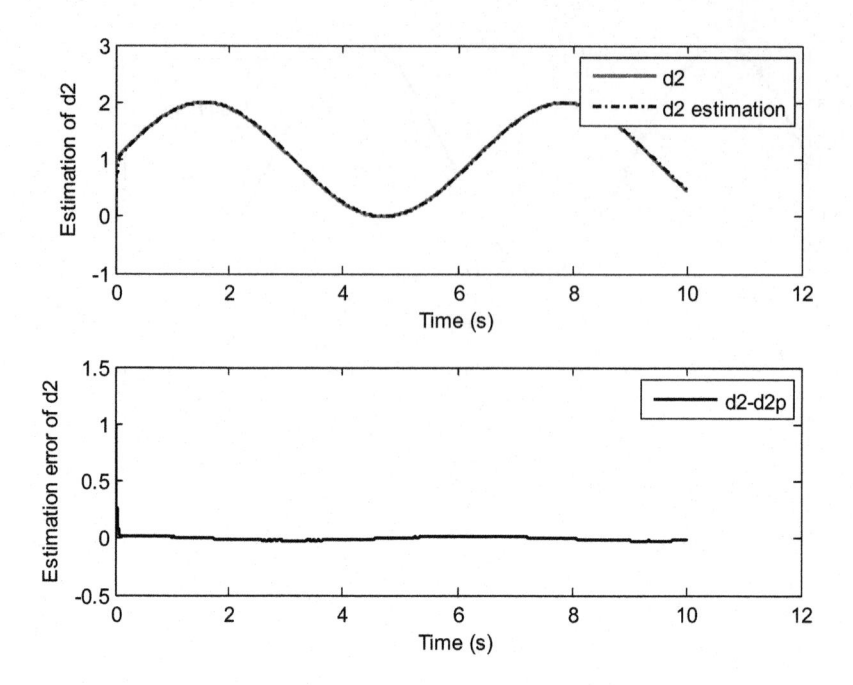

Fig. 10.3 Estimation of d_2

10.3 Simulation Example

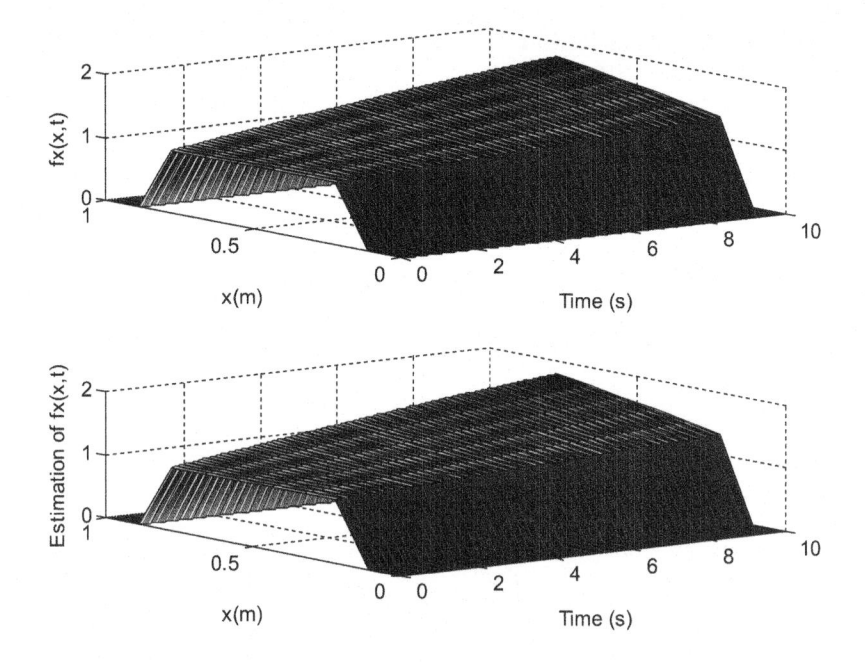

Fig. 10.4 Estimation of $f(x)$

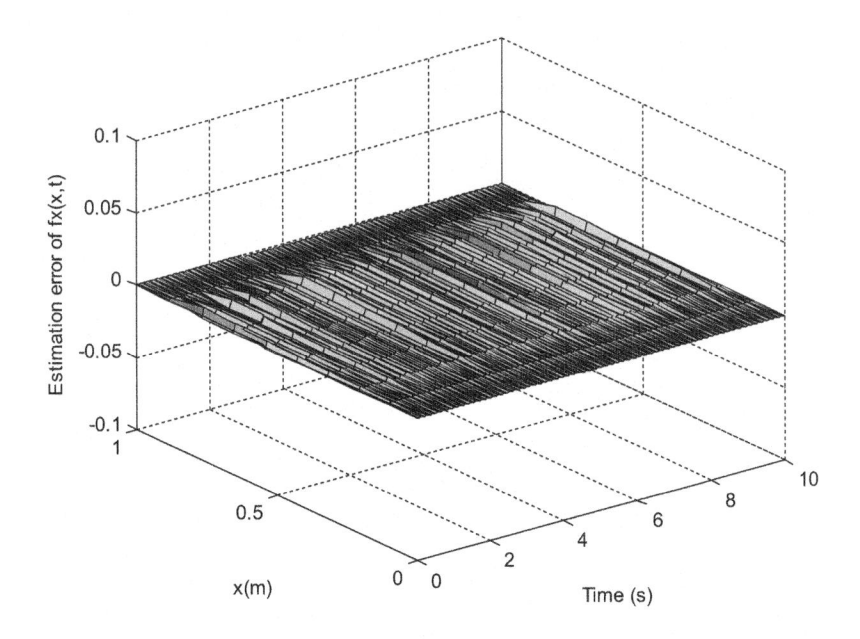

Fig. 10.5 Estimation error of $f(x)$

Simulation program: chap10_1.m

```
close all;
clear all;
nx=10;
nt=20000;

tmax=10;L=1;
%Compute mesh spacing and time step
dx=L/(nx-1);
T=tmax/(nt-1);
%Create arrays to save data for export
t=linspace(0,nt*T,nt);
x=linspace(0,L,nx);
%Parameters
EI=3;rho=0.2;m=0.1;Ih=0.1;
L1=50;L2=50;L3=1000;
%Define viriables and Initial condition:
y=zeros(nx,nt);
dy=zeros(nx,nt);
fx=zeros(nx,nt);
fxp=zeros(nx,nt);
w3p=zeros(nx,nt);
z=zeros(nx,nt);     %elastic deflectgion
for j=1:nt
    th(j)=0;        %joint angle
    d1p(j)=0;d2p(j)=0;
    w1p(j)=0;w2p(j)=0;
    dzL(j)=0;
end

for j=3:nt          %Begin
tol(j)=sin(j*T);
F(j)=sin(j*T);
d1(j)=1+sin(j*T);
d2(j)=1+sin(j*T);
for i=3:nx-2
    fx(i,j)=1+1*sin(0.1*i*dx*j*T);
end
yxx0=(y(3,j-1)-2*y(2,j-1)+y(1,j-1))/dx^2;
zxx0=yxx0;
th(j)=2*th(j-1)-th(j-2)+T^2/Ih*(tol(j)+EI*zxx0+d1(j-1));    %(10.2)


%%%%%%%%%%%%%%%%%%%%%%%%%%%%%%%%%%%%%%%%%%%%%%%%%%%%%%%%%%%%
%z(i,j)
dth(j)=(th(j)-th(j-1))/T;
ddth(j)=(dth(j)-dth(j-1))/T;

%get z(i,j),i=1,2, Boundary condition
y(1,:)=0;       %y(0,t)=0, i=1
y(2,:)=0;       %y(1,t)=0, i=2
z(1,:)=0;       %y(0,t)=0, i=1
z(2,:)=0;       %y(1,t)=0, i=2

%get y(i,j),i=3:nx-2
for i=3:nx-2
    yxxxx=(y(i+2,j-1)-4*y(i+1,j-1)+6*y(i,j-1)-4*y(i-1,j-1)+y(i-2,j-1))/dx^4;
    y(i,j)=T^2*(-i*dx*ddth(j-1)-(EI*yxxxx-fx(i,j-1))/rho)+2*y(i,j-1)-y(i,j-2);  %(10.1)
    zxxxx(i,j-1)=yxxxx;%i*dx=x

    dy(i,j)=(y(i,j)-y(i,j-1))/T;
    dzx(i,j)=i*dx*dth(j)+dy(i,j);
end

%get z(nx-1,j),i=nx-1
yxxxx(nx-1,j-1)=(-2*y(nx,j-1)+5*y(nx-1,j-1)-4*y(nx-2,j-1)+y(nx-3,j-1))/dx^4;
y(nx-1,j)=T^2*(-(nx-1)*dx*ddth(j-1)-(EI*yxxxx(nx-1,j-1)-fx(nx-1,j-1))/rho)+2*y(nx-1,j-1)-y(nx-1,j-2);  %(10.1)
zxxxx(nx-1,j-1)=yxxxx(nx-1,j-1);
dy(nx-1,j)=(y(nx-1,j)-y(nx-1,j-1))/T;

%get y(nx,j),y=nx
yxxxL(j-1)=(-y(nx,j-1)+2*y(nx-1,j-1)-y(nx-2,j-1))/dx^3;
zxxxL(j-1)=yxxxL(j-1);
dzxxx_L=(yxxxL(j-1)-yxxxL(j-2))/T;
```

10.3 Simulation Example

```
y(nx,j)=T^2*(-L*ddth(j-1)+(EI*yxxxL(j-1)+F(j-1)+d2(j-1))/m)+2*y(nx,j-1)-y(nx,j-2);
%(10.3)

dyL(j)=(y(nx,j)-y(nx,j-1))/T;
dy(nx,j)=dyL(j);
dzL(j)=L*dth(j)+dyL(j);

w1p(j)=w1p(j-1)+T*(L1*(-EI*zxx0-tol(j-1))-L1*d1p(j-1));   %(10.5)
P1(j)=L1*Ih*dth(j);
d1p(j)=w1p(j)+P1(j);
w2p(j)=w2p(j-1)+T*(L2*(-EI*zxxxL(j-1)-F(j-1))-L2*d2p(j-1));    %(10.6)
P2(j)=L2*m*dzL(j);
d2p(j)=w2p(j)+P2(j);

end     %End
%To view the curve, short the points
  for j=3:nt
    for i=3:nx-2
    w3p(i,j)=w3p(i,j-1)+T*(L3*EI*zxxxx(i,j-1)-L3*fxp(i,j-1));    %(10.7)
    P3(i,j)=L3*rho*dzx(i,j);
    fxp(i,j)=w3p(i,j)+P3(i,j);
      end
end

tshort=linspace(0,tmax,nt/100);
fxshort=zeros(nx,nt/100);
fxpshort=zeros(nx,nt/100);
for j=1:nt/100
      for i=1:nx
          fxshort(i,j)=fx(i,j*100);
          fxpshort(i,j)=fxp(i,j*100);
      end
end
yshort=zeros(nx,nt/100);
dyshort=zeros(nx,nt/100);
for j=1:nt/100
      for i=1:nx
          yshort(i,j)=y(i,j*100);
          dyshort(i,j)=dy(i,j*100);
      end
end
figure(1);
subplot(211);
plot(t,d1,'r',t,d1p,'-.k','linewidth',2);
xlabel('Time (s)');ylabel('Estimation of d1');
legend('d1', 'd1 estimation');
subplot(212);
plot(t,d1-d1p,'k','linewidth',2);
xlabel('Time (s)');ylabel('Estimation error of d1');
legend('d1-d1p');

figure(2);
subplot(211);
plot(t,d2,'r',t,d2p,'-.k','linewidth',2);
xlabel('Time (s)');ylabel('Estimation of d2');
legend('d2', 'd2 estimation');
subplot(212);
plot(t,d2-d2p,'k','linewidth',2);
xlabel('Time (s)');ylabel('Estimation error of d2');
legend('d2-d2p');

figure(3);
subplot(211);
surf(tshort,x,fxshort);
xlabel('Time (s)'); ylabel('x(m)');zlabel('fx(x,t)');
subplot(212);
surf(tshort,x,fxpshort);
xlabel('Time (s)'); ylabel('x(m)');zlabel('Estimation of fx(x,t)');

figure(4);
surf(tshort,x,fxshort-fxpshort);
xlabel('Time (s)'); ylabel('x(m)');zlabel('Estimation error of fx(x,t)');
axis([0 10 0 1 -0.1 0.1]);
```

References

1. S.S. Ge, S. Zhang, W. He, Vibration control of an Euler-Bernoulli beam under unknown spatiotemporally varying disturbance. Int. J. Control **84**(5), 947–960 (2011)
2. T.T. Jiang, J.K. Liu, W. He, Boundary control for a flexible manipulator based on infinite dimensional disturbance observer. J. Sound Vib. **348**(21), 1–14 (2015)

Chapter 11
Boundary Control for Flexible Manipulator with Guaranteed Transient Performance

Recently, a novel prescribed performance constraint method was developed in [1–3]. In these studies, using a prescribed error transformation function, a prescribed performance can be guaranteed, which means that the tracking performance of the transient and steady-state error can be regulated. Several applications have been devised using this technique.

In [4], the authors use a prescribed error transformation function technique to solve the boundary control problem of a flexible manipulator based on the PDE model with input disturbances and output constraints we introduce a basic boundary controller design method for a flexible manipulator with guaranteed transient performance.

11.1 System Description

Considering PDE model given in Chap. 3, neglecting disturbance, the PDE model is given as

$$\rho\left(x\ddot{\theta}+\ddot{y}(x)\right) = -EIy_{xxxx}(x) \tag{11.1}$$

$$\tau = I_h\ddot{\theta} - EIy_{xx}(0) \tag{11.2}$$

$$F = m\left(L\ddot{\theta}+\ddot{y}(L)\right) - EIy_{xxx}(L) \tag{11.3}$$

$$y(0) = 0,\ y_x(0) = 0,\ y_{xx}(L) = 0 \tag{11.4}$$

Define $z(x) = x\theta + y(x)$, then $\ddot{z}(x) = x\ddot{\theta}+\ddot{y}(x)$, $\ddot{z}(L) = L\ddot{\theta}+\ddot{y}(L)$. From $z(x)$ definition, we have

$$z_{xx}(x) = y_{xx}(x),\ \ddot{z}_x(0) = \ddot{\theta},\ z_{xx}(0) = y_{xx}(0),\ z_{xx}(L) = y_{xx}(L),\ z_{xxx}(L) = y_{xxx}(L) \tag{11.5}$$

Considering $\theta_\mathrm{d}(t)$ as ideal angle signal, define θ_d as constant value.

11.2 Preliminaries

This part presents the following necessary Lemmas, which will be needed in the subsequent design and analysis of boundary control scheme.

Lemma 1 [5] *For $\phi_1(x,t)$, $\phi_2(x,t) \in R$, $x \in [0, L]$, $t \in [0, \infty)$, there exists*

$$\begin{aligned}\phi_1(x,t)\phi_2(x,t) &\le |\phi_1(x,t)\phi_2(x,t)| \le \phi_1^2(x,t) + \phi_2^2(x,t) \\ |\phi_1(x,t)\phi_2(x,t)| &\le \frac{1}{\gamma}\phi_1^2(x,t) + \gamma\phi_2^2(x,t)\end{aligned} \tag{11.6}$$

where $\gamma > 0$.

Lemma 2 [6] *For $\phi(x,t) \in R$, $x \in [0, L]$, $t \in [0, \infty)$, if $p(0,t) = 0, \forall t \in [0, \infty)$, then*

$$\phi^2(x,t) \le L\int_0^L \phi_x^2(x,t)\,dx,\ \forall x \in [0, L] \tag{11.7}$$

Similarly, if $\phi_x(0,t) = 0, \forall t \in [0, \infty)$, then

$$\phi_x^2(x,t) \le L\int_0^L \phi_{xx}^2(x,t)\,dx,\ \forall x \in [0, L] \tag{11.8}$$

Lemma 3 *For $f(x) \le g(x), x \in [a,b]$, then*

$$\int_a^b f(x)\,dx \le \int_a^b g(x)\,dx \tag{11.9}$$

The control objectives are:

(1) To design boundary control for regulating angle and suppressing elastic vibration simultaneously while ensuring that all signals of the closed-loop systems are bounded.
(2) The prescribed performance bounds on the angle tracking error $e_1 = \theta - \theta_\mathrm{d}$ and the deflection at the end $e_2 = y(L,t)$ are satisfied at any time.

11.3 Performance Function

To achieve the second objective, a performance function is selected, where a smooth function $\lambda: R_+ \to R_+$, originally appeared in [3], will be called a performance function if: (1) $\lambda(t)$ is positive and decreasing; (2) $\lim\limits_{t \to \infty} \lambda(t) = \lambda_\infty > 0$.

Then the transient performances are guaranteed by the following prescribed constraint conditions:

$$-\lambda_i(t) < e_i(t) < \lambda_i(t), \quad i = 1, 2 \tag{11.10}$$

For all $t \geq 0$. The exponentially decaying performance functions are selected in the following form

$$\lambda_i(t) = (\lambda_i(0) - \lambda_{i\infty}) \exp(-lt) + \lambda_{i\infty}$$

where λ_{i0}, $\lambda_{i\infty}$ and l are appropriately defined positive constants and $\lambda_{i0} > \lambda_{i\infty}$.

Remark Select $0 < |e_i(0)| < \lambda_i(0)$. Moreover, the constant $\lambda_{i\infty}$ represents the maximum allowable value if the errors are at the steady state. The decreasing rate of $\lambda_i(t)$ introduces a lower bound on the required speed of convergence of $e_i(t)$, while the maximum overshoot is prescribed less than $\lambda_i(0)$.

11.4 Controller Design and Analysis

Define

$$e_i(t) = \lambda_i(t) S_i(\varepsilon_i), \quad i = 1, 2 \tag{11.11}$$

where ε_i is the transformed error and $S_i(\varepsilon_i)$ is a function defined as

$$S_i(\varepsilon_i) = \tanh(\varepsilon_i) = \frac{e^{\varepsilon_i} - e^{-\varepsilon_i}}{e^{\varepsilon_i} + e^{-\varepsilon_i}} \tag{11.12}$$

Obviously, we have

$$-1 < S_i(\varepsilon_i) < 1 \tag{11.13}$$

Hence, with the function $S_i(\varepsilon_i)$, (11.11) gives

$$-\lambda_i(t) < e_i(t) < \lambda_i(t), \quad i = 1, 2 \tag{11.14}$$

Furthermore, since the function $S_i(\varepsilon_i)$ is strictly monotonic increasing, its inverse function is

$$\varepsilon_i = \frac{1}{2}\ln\frac{1+S_i}{1-S_i} = \frac{1}{2}\ln\frac{1+\frac{e_i}{\lambda_i}}{1-\frac{e_i}{\lambda_i}} = \frac{1}{2}\ln\frac{\lambda_i+e_i}{\lambda_i-e_i} = \frac{1}{2}(\ln(\lambda_i+e_i) - \ln(\lambda_i-e_i)) \quad (11.15)$$

Differentiating (11.15) with respect to time, we get

$$\dot{\varepsilon}_i = \frac{1}{2}\left(\frac{\dot{\lambda}_i+\dot{e}_i}{\lambda_i+e_i} - \frac{\dot{\lambda}_i-\dot{e}_i}{\lambda_i-e_i}\right) = \frac{\lambda_i\dot{e}_i - \dot{\lambda}_i e_i}{\lambda_i^2 - e_i^2} \quad (11.16)$$

where $i = 1, 2$.

Clearly, owing to the properties of error transformation, (11.10) is guaranteed if we are able to keep ε_i, $i = 1, 2$ bounded $\forall t \geq 0$.

To guarantee that the tracking error $e_1 = \theta - \theta_d$ and the deflection at the end $e_2 = y(L,t)$ tend to zero, and ε_i, $i = 1, 2$ is bounded, we define the control law as

$$F = \alpha_2 e_2 - \frac{\alpha_4 \varepsilon_2 \lambda_2}{\lambda_2^2 - e_2^2} - \alpha_6 \dot{e}_2 \quad (11.17)$$

$$\tau = \alpha_1 e_1 - \frac{\alpha_3 \varepsilon_1 \lambda_1}{\lambda_1^2 - e_1^2} - \alpha_5 \dot{e}_1 - LF \quad (11.18)$$

where $\alpha_1, \alpha_2, \alpha_3, \alpha_4, \alpha_5$ and, α_6 are positive constants, and are designed in the following process of proof.

Theorem 11.1 [4] Considering the system (11.1)–(11.4), using the proposed control laws (11.17) and (11.18), the following properties hold:

(1) The closed-loop system is asymptotically stable, that is $\theta(t) \to \theta_d(t)$, $y(x) \to 0$;
(2) If the initial boundary conditions $|e_i(t)| < \lambda_i(0)$, $i = 1, 2$, then the tracking error $e_1(t)$ remains in the set $\Omega_{e_1} := \{e_1(t) \in R : |e_1(t)| < \lambda_1(0)\}$ and the deflection at the end $e_2(t) = y(L,t)$ remains in the set $\Omega_{e_2} := \{e_2(t) \in R : |e_2(t)| < \lambda_2(0)\}$. Moreover, the prescribed transient and steady state error bounds are achieved.

Proof Lyapunov candidate function is taken to be

$$V = V_1 + V_2 + V_3$$

where

$$V_1 = \frac{1}{2}\int_0^L \rho \dot{z}^2(x)\,dx + \frac{1}{2}EI\int_0^L y_{xx}^2(x)\,dx$$

$$V_2 = \frac{1}{2}I_h\dot{e}_1^2 + \frac{1}{2}m\dot{z}^2(L,t) + \frac{1}{2}\alpha_1 e_1^2 + \frac{1}{2}\alpha_2 e_2^2 + \frac{1}{2}\alpha_3 \varepsilon_1^2 + \frac{1}{2}\alpha_4 \varepsilon_2^2$$

$$V_3 = \beta I_h e_1 \dot{e}_1 + \beta m e_2 \dot{z}(L) + \beta L m e_1 \dot{z}(L) + \beta \rho \int_0^L x\dot{z}(x)(e_1 + y_x(x))\,dx$$

11.4 Controller Design and Analysis

Since $\dot{z}(x)(e_1 + y_x(x)) = \dot{z}(x)e_1 + \dot{z}(x)y_x(x) \leq e_1^2 + y_x^2(x) + 2\dot{z}^2(x)$, according to Lemmas 11.1–11.3, we have

$$\int_0^L x\dot{z}(x)(e_1 + y_x(x))\,dx \leq \int_0^L x\left(e_1^2 + y_x^2(x) + 2\dot{z}^2(x)\right) dx$$

$$= Le_1^2 + L^3 \int_0^L y_{xx}^2(x)\,dx + 2L \int_0^L \dot{z}^2(x)\,dx$$

Then we have

$$|V_3| \leq \frac{1}{\alpha_1}\beta(I_h + Lm + \rho L)\frac{\alpha_1}{2}e_1^2 + \beta\frac{I_h}{2}\dot{e}_1^2 + + \frac{\beta m}{\alpha_2}\frac{\alpha_2}{2}e_2^2 + \frac{\beta + \beta L}{2}m\dot{z}^2(L) + \frac{\beta Lm}{2}m\dot{z}^2(L)$$

$$+ 2\beta L\frac{\rho}{2}\int_0^L \dot{z}^2(x)\,dx + \beta\rho L^3\frac{EI}{2EI}\int_0^L y_{xx}^2(x)\,dx \leq \delta(V_1 + V_2)$$

where

$$\delta = \max\left\{\frac{\beta(I_h + Lm + \rho L)}{\alpha_1}, \frac{\beta m}{\alpha_2}, \beta + \beta L, 2\beta L, \frac{\beta\rho L^3}{EI}\right\} \quad (11.19)$$

where β is a positive constant. It is obvious that V is positive define when β is chosen to satisfy the inequality $0 < \delta < 1$.

Then the derivative of V with respect to time is

$$\dot{V}_1 = \int_0^L \rho\dot{z}(x,t)\ddot{z}(x,t)\,dx + EI\int_0^L y_{xx}(x,t)\dot{y}_{xx}(x,t)\,dx$$

$$= -EIy_{xxx}(L,t)L\dot{\theta}(t) - EIy_{xx}(0,t)\dot{\theta}(t) - EIy_{xxx}(L,t)\dot{y}(L,t)$$

$$= -EIy_{xxx}(L,t)\dot{z}(L,t) - EIy_{xx}(0,t)\dot{\theta}(t)$$

$$\dot{V}_2 = \dot{e}_1\left(\tau + EIy_{xx}(0,t) + \alpha_1 e_1 + \frac{\alpha_3\varepsilon_1\lambda_1}{\lambda_1^2 - e_1^2} + LF\right) - \alpha_3\varepsilon_1\frac{\dot{\lambda}_1 e_1}{\lambda_1^2 - e_1^2} - \alpha_4\varepsilon_2\frac{\dot{\lambda}_2 e_2}{\lambda_2^2 - e_2^2}$$

$$+ \dot{e}_2\left(F + \alpha_2 e_2 + \frac{\alpha_4\varepsilon_2\lambda_2}{\lambda_2^2 - e_2^2}\right) + \dot{z}(L,t)EIy_{xxx}(L,t)$$

$$\dot{V}_3 = \beta I_h \dot{e}_1^2 + \beta e_1(\tau + EIy_{xx}(0,t)) + \beta m\dot{e}_2 \dot{z}(L,t) + \beta e_2(F + EIy_{xxx}(L,t)) + \beta Lm\dot{e}_1 \dot{z}(L,t)$$
$$+ \beta L e_1(F + EIy_{xxx}(L,t)) - EI\beta L e_1 y_{xxx}(L,t) - EI\beta e_2 y_{xxx}(L,t)$$
$$- \frac{3}{2}\beta EI \int_0^L y_{xx}^2(x,t)dx - EI\beta e_1 y_{xx}(0,t)$$
$$+ \frac{1}{2}\beta \rho L \dot{z}^2(L,t) - \beta\rho \int_0^L \dot{z}^2(x,t)\,dx$$

Using (11.2), (11.3), (11.17) and (11.18), we have

$$\dot{V} = -\alpha_5 \dot{e}_1^2 - \alpha_6 \dot{e}_2^2 - \left(\dot{\lambda}_1 + \beta\lambda_1\right)\frac{\alpha_3 \varepsilon_1 e_1}{\lambda_1^2 - e_1^2} - \left(\dot{\lambda}_2 + \beta\lambda_2\right)\frac{\alpha_4 \varepsilon_2 e_2}{\lambda_2^2 - e_2^2}$$
$$+ \beta I_h \dot{e}_1^2 - \beta\alpha_1 e_1^2 - \beta\alpha_5 e_1 \dot{e}_1 + \beta m\dot{e}_2 \dot{z}(L,t)$$
$$- \beta\alpha_2 e_2^2 - \beta\alpha_6 e_2 \dot{e}_2 + \beta Lm\dot{e}_1 \dot{z}(L,t) - \frac{3}{2}\beta EI \int_0^L y_{xx}^2(x,t)\,dx + \frac{1}{2}\beta\rho L\dot{z}^2(L,t) \quad (11.20)$$
$$- \beta\rho \int_0^L \dot{z}^2(x,t)\,dx$$

Using Lemmas 11.1 to 11.3, considering $z(x) = x\theta + y(x)$, we have $z(L) = L\theta + y(L)$ and $\dot{z}(L) = L\dot{e}_1 + \dot{e}_2$, $\dot{z}^2(L) = L^2 \dot{e}_1^2 + \dot{e}_2^2 + 2L\dot{e}_1\dot{e}_2 \leq L^2\dot{e}_1^2 + \dot{e}_2^2 + L(\dot{e}_1^2 + \dot{e}_2^2)$, then

$$\dot{e}_2 \dot{z}(L,t) = \dot{e}_2(L\dot{e}_1 + \dot{e}_2) \leq \frac{L}{2}\dot{e}_1^2 + \frac{L}{2}\dot{e}_2^2 + \dot{e}_2^2$$
$$\dot{e}_1 \dot{z}(L,t) = \dot{e}_1(L\dot{e}_1 + \dot{e}_2) = L\dot{e}_1^2 + \dot{e}_1\dot{e}_2 \leq L\dot{e}_1^2 + \frac{1}{2}\dot{e}_1^2 + \frac{1}{2}\dot{e}_2^2$$

Then we can get the following five inequalities

(1) $-\beta\alpha_5 e_1 \dot{e}_1 \leq \frac{\beta\alpha_5}{2\delta_1} e_1^2 + \frac{\beta\alpha_5 \delta_1}{2}\dot{e}_1^2$;
(2) $\beta m\dot{e}_2 \dot{z}(L,t) \leq \frac{\beta mL}{2}\dot{e}_1^2 + \frac{\beta mL}{2}\dot{e}_2^2 + \beta m\dot{e}_2^2$;
(3) $\beta\alpha_6 e_2 \dot{e}_2 \leq \frac{\beta\alpha_6}{2\delta_2} e_2^2 + \frac{\beta\alpha_6 \delta_2}{2}\dot{e}_2^2$;
(4) $\beta Lm\dot{e}_1 \dot{z}(L,t) \leq \beta Lm L\dot{e}_1^2 + \frac{\beta Lm}{2}\dot{e}_1^2 + \frac{\beta Lm}{2}\dot{e}_2^2$;
(5) $\frac{1}{2}\dot{z}^2(L,t) \leq \frac{1}{2}(L^2 + L)\dot{e}_1^2 + \frac{1}{2}(1+L)\dot{e}_2^2$.

where $\delta_i\,(i = 1,2)$ are positive constants.

11.4 Controller Design and Analysis

Equation (11.20) can be rewritten as

$$|V_3| \leq -\left(\alpha_5 - \beta I_h - \beta mL - \frac{\beta\alpha_5\delta_1}{2} - \beta mL^2\right)\dot{e}_1^2 + \frac{1}{2}\beta\rho(L^3+L^2)\dot{e}_1^2 - \left(\beta\alpha_1 - \frac{\beta\alpha_5}{2\delta_1}\right)e_1^2$$
$$-\left(\alpha_6 - \beta mL - \beta m - \frac{\beta\alpha_6\delta_2}{2} - \frac{1}{2}\beta\rho(L+L^2)\right)\dot{e}_2^2 - \left(\beta\alpha_2 - \frac{\beta\alpha_6}{2\delta_2}\right)e_2^2 - \left(\dot{\lambda}_1 + \beta\lambda_1\right)\frac{\alpha_3\varepsilon_1 e_1}{\lambda_1^2 - e_1^2}$$
$$-\left(\dot{\lambda}_2 + \beta\lambda_2\right)\frac{\alpha_4\varepsilon_2 e_2}{\lambda_2^2 - e_2^2} - \frac{3}{2}\beta EI\int_0^L y_{xx}^2(x,t)\,dx - \beta\rho\int_0^L \dot{z}^2(x,t)\,dx$$

We design parameters α_1, α_2, α_5 and, α_6. δ_1, δ_2 and β to satisfy the following inequalities

$$\gamma_1 = \alpha_5 - \beta I_h - \beta mL - \frac{\beta\alpha_5\delta_1}{2} - \beta mL^2 - \frac{1}{2}\beta\rho(L^3+L^2) > 0 \quad (11.21a)$$

$$\gamma_2 = \alpha_6 - \beta mL - \beta m - \frac{\beta\alpha_6\delta_2}{2} - \frac{1}{2}\beta\rho(L+L^2) > 0 \quad (11.21b)$$

$$\gamma_3 = \beta\alpha_1 - \frac{\beta\alpha_5}{2\delta_1} > 0 \quad (11.21c)$$

$$\gamma_4 = \beta\alpha_2 - \frac{\beta\alpha_6}{2\delta_2} > 0 \quad (11.21d)$$

$$\gamma_5 = \dot{\lambda}_1 + \beta\lambda_1 \geq 0 \quad (11.21e)$$

$$\gamma_6 = \dot{\lambda}_2 + \beta\lambda_2 \geq 0 \quad (11.21f)$$

Noting that $\varepsilon_i \frac{e_i}{\lambda_i^2 - e_i^2} \geq 0$, $i = 1, 2$, we have

$$\dot{V} \leq -\gamma_1\dot{e}_1^2 - \gamma_2\dot{e}_2^2 - \gamma_3 e_1^2 - \gamma_4 e_2^2 \leq 0 \quad (11.22)$$

Considering the same PDE model, in this chapter, the dissipative and unique analysis are similar to the analysis of Chap. 8.

11.5 Convergence Analysis

Observing the right-hand of (11.22), it follows that, $\dot{V}$ is negative semi-definite and $\dot{V} \equiv 0$ only if $e_1 \equiv e_2 \equiv \dot{e}_1 \equiv \dot{e}_2 \equiv 0$. We further obtain that $\ddot{e}_1 \equiv \ddot{e}_2 \equiv 0$. That is

$$\ddot{\theta} \equiv \ddot{y}(L,t) \equiv 0 \quad (11.23)$$

Applying (11.23) in (11.1), we have

$$\rho \ddot{y}(x,t) = -EI y_{xxxx}(x,t) \tag{11.24}$$

$$y_{xxxx}(L,t) \equiv 0 \tag{11.25}$$

To go on, (11.24) is separable and can be treated by the technique of separation of variables [7]. We assume that $y(x,t)$ can be written as follows:

$$y(x,t) = W(x) \cdot \phi(t) \tag{11.26}$$

where $W(x,t)$ and $\phi(t)$ are unknown functions of space and time to be determined. Based on (11.24) and (11.26), it yields

$$y_{xxxx}(x,t) = -\frac{\rho}{EI}\ddot{y}(x,t)$$

From (11.26), we have $y_{xxxx}(x,t) = W^{(4)}(x) \cdot \phi(t)$, $\ddot{y}(x,t) = W(x) \cdot \phi''(t)$, then above equation becomes

$$\frac{W^{(4)}(x)}{W(x)} = -\frac{\rho}{EI}\frac{\phi''(t)}{\phi(t)} = \mu \tag{11.27}$$

where $\phi''(t) = \frac{d^2\phi(t)}{dt^2}$, $W^{(4)}(x) = \frac{d^4 W}{dx^4}$.
Then we have

$$W^{(4)}(x) - \mu W(x) = 0 \tag{11.28}$$

$$\phi''(t) + \frac{EI\mu}{\rho}\phi(t) = 0 \tag{11.29}$$

let $\mu = \eta^4$, then we can get the solution (11.28) as

$$W(x) = c_1 \cosh \eta x + c_2 \sinh \eta x + c_3 \cos \eta x + c_4 \sin \eta x \tag{11.30}$$

where $c_i \in R$, $i = 1,2,3,4$ are unknown real number to be determined.
Using (11.4), (11.25) and (11.26), we have $W(0) = W'(0) = W''(L) = W^{(4)}(L) = 0$, then from (11.30), we have

$$\begin{cases} c_1 + c_3 = 0 \\ c_2 + c_4 = 0 \\ c_1 \cosh \beta L + c_2 \sinh \beta L - c_3 \cos \beta L - c_4 \sin \beta L = 0 \\ c_1 \cosh \beta L + c_2 \sinh \beta L + c_3 \cos \beta L + c_4 \sin \beta L = 0 \end{cases} \tag{11.31}$$

then we have

$$\begin{cases} c_1 \cosh \beta L + c_2 \sinh \beta L = 0 \\ c_3 \cos \beta L + c_4 \sin \beta L = 0 \end{cases}$$

i.e.

$$\begin{cases} c_3 \cosh \beta L + c_4 \sinh \beta L = 0 \\ c_3 \cos \beta L + c_4 \sin \beta L = 0 \end{cases}$$

therefore

$$c_4(\sinh \beta L \cdot \cos \beta L - \sin \beta L \cdot \cosh \beta L) = 0$$

We can conclude that $W^{(4)}(x) - \mu W(x) = 0$ have unique solutions, $c_i = 0$, $i = 1, 2, 3, 4$, thus, $W(x) = 0$, $y(x,t) = 0$.

Considering if $\dot{V} \equiv 0$, $e_1 \equiv e_2 \equiv \dot{e}_1 \equiv \dot{e}_2 \equiv 0$, according to the extended LaSalle's invariance principle [8], the PDE boundary control (11.17) and (11.18) can guarantee the asymptotic stability of the closed-loop system, thus $y(x,t) \to 0$, and $\theta(t) \to \theta_d(t)$, $\dot{\theta}(t) \to \dot{\theta}_d(t)$, $\dot{y}(L,t) \to 0$, errors $e_i(t)$, $i = 1, 2$ will tend to be in the prescribed performance bounds.

11.6 Simulation Example

Considering the PDE model as Eqs. (11.1)–(11.4), the physical parameters are chosen as: $EI = 2.0$, $L = 1.0$, $\rho = 0.20$, $m = 0.20$, $I_h = 0.50$.

Define ideal angle as $\theta_d = 0.5$, use controller (11.17) and (11.18). We use chap11_1.m to test the inequalities in (11.19) and (11.21a)–(11.21d), and choose $\alpha_1 = \alpha_2 = \alpha_3 = \alpha_4 = \alpha_5 = \alpha_6 = 10$.

For the performance function, to satisfy (11.21e)–(11.21f), we set small l and choose $l = 0.10$, set $\lambda_{10} = 0.55$, $\lambda_{1\infty} = 0.01$, $\lambda_{20} = 0.12$, $\lambda_{2\infty} = 0.01$. Two axes are divided into sections as $nx = 9$, $nt = 20001$. The simulation results are shown from Figs. 11.1, 11.2, 11.3, 11.4, 11.5 and 11.6.

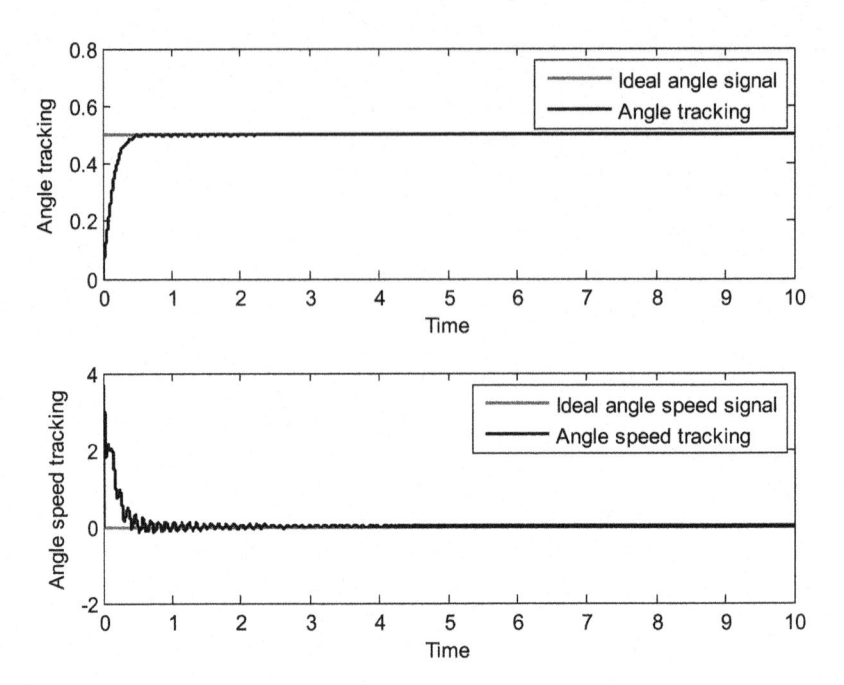

Fig. 11.1 Angle tracking and angle speed tracking

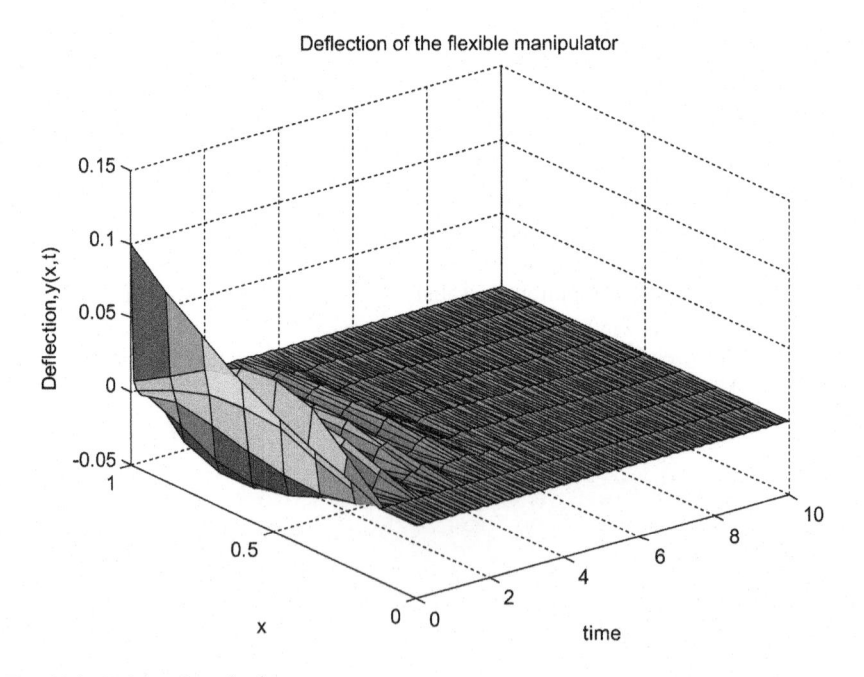

Fig. 11.2 Deformation $y(x, t)$

11.6 Simulation Example

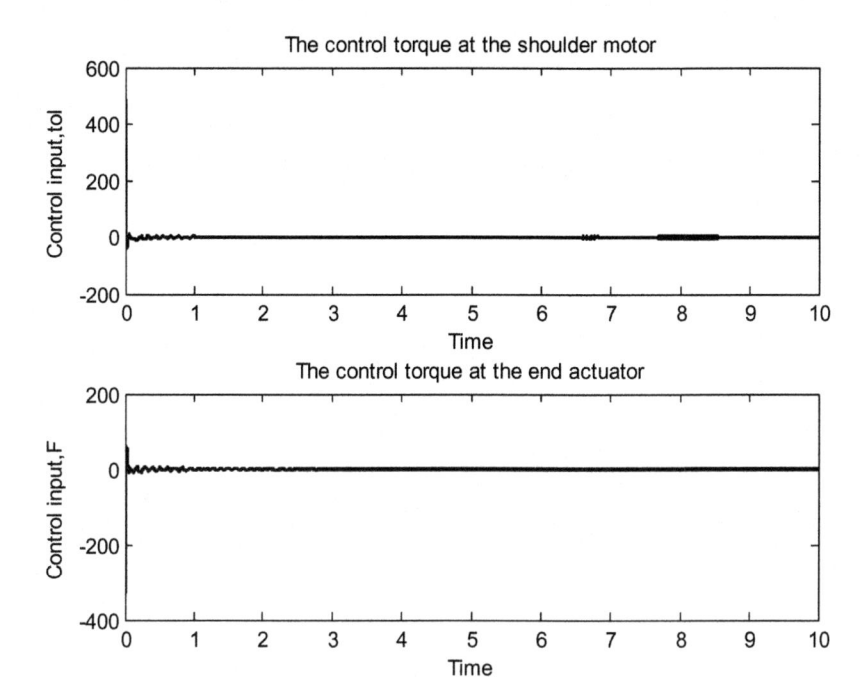

Fig. 11.3 Boundary control input, τ and F

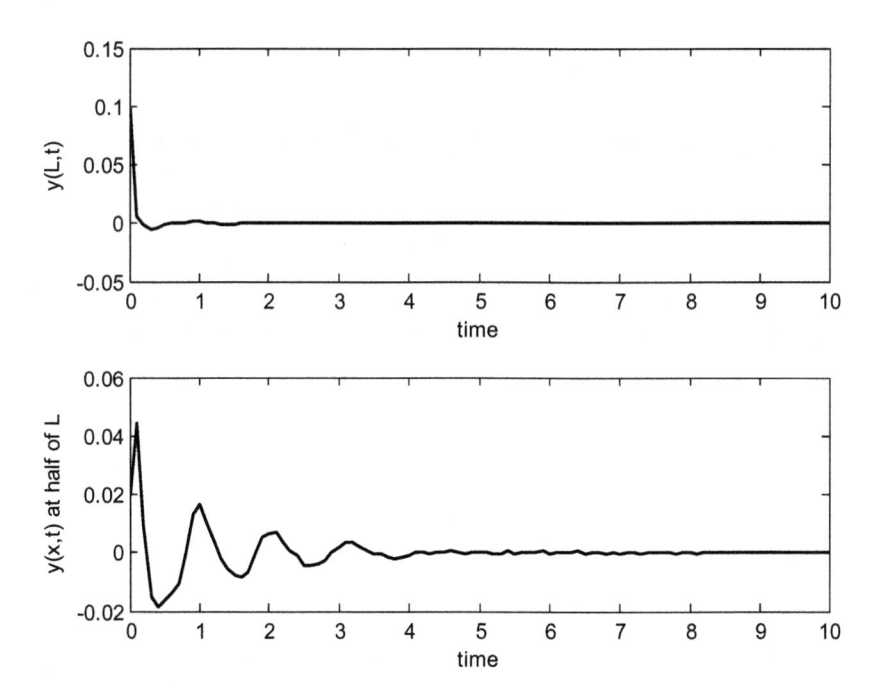

Fig. 11.4 Deformation at $x = \frac{L}{2}$ and $x = L$

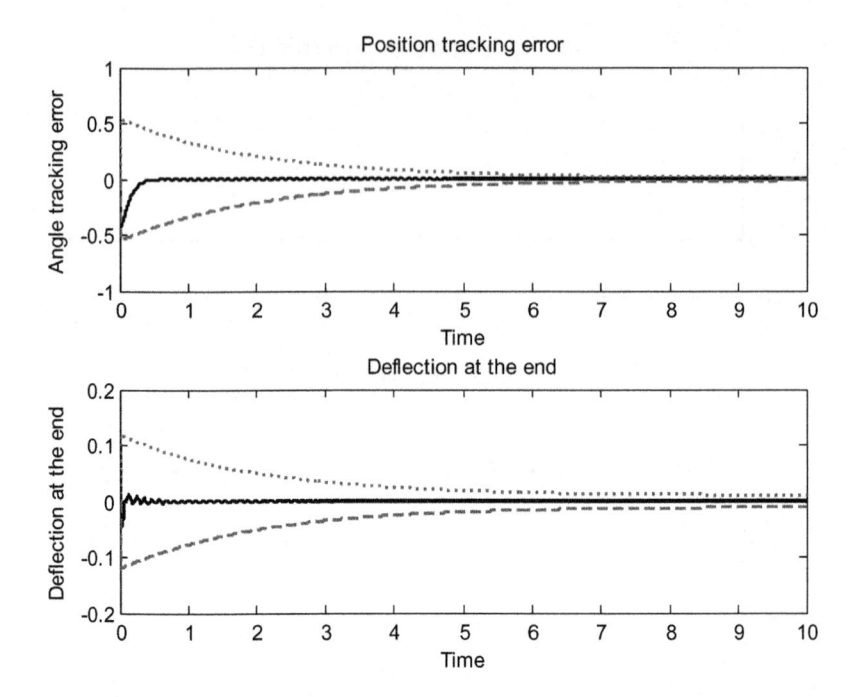

Fig. 11.5 The tracking error and deflection at the end

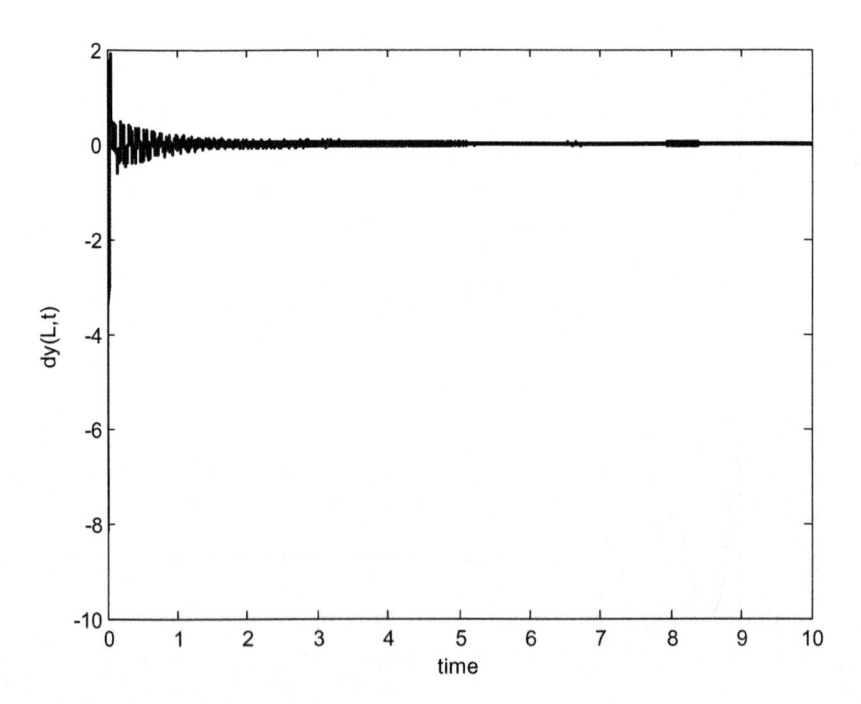

Fig. 11.6 Deformation rate $\dot{y}(L, t)$

11.6 Simulation Example

Simulation program:

(1) **Test program in (11.19) and (11.21): chap11_1.m**

```
close all;
clear all;
%Parameters
EI=2;rho=0.2;m=0.2;Ih=0.5;

L=1;l=0.10;
lamda10=0.55;lamda1_inf=0.01;
lamda20=0.12;lamda2_inf=0.01;

alfa1=10;alfa2=10;alfa3=10;alfa4=10;alfa5=10;alfa6=10;
beta=0.49;

X=[beta*(Ih+L*m+rho*L)/alfa1,
beta*m/alfa2,beta+beta*L,2*beta*L,beta*rho*L^3/EI];
delta=max(X)
%%%%%%%%%%%%%%%%%%%%%%%%%%%%%%%%%%%%%%%%%%%%%%%%%%%%%%%%%%%%%%%%
%%%%%%
delta1=1.0;
delta2=1.0;
gama1=alfa5-beta*Ih-beta*m*L-beta*alfa5*delta1-beta*L^2-0.5*beta*rho*
(L^3+L^2)
gama2=alfa6-beta*m*L-beta*m-beta*alfa6*delta2-0.5*beta*rho*(L+L^2)
gama3=beta*alfa1-beta*alfa5/(2*delta1)
gama4=beta*alfa2-beta*alfa6/(2*delta2)
```

(2) Main program: chap11_2.m

```
close all;
clear all;

nx=8+1;nt=10000+1;
tmax=10.0;L=1;
%Compute mesh spacing and time step
dx=L/(nx-1);
T=tmax/(nt-1);

%Create arrays to save data for export
t=linspace(0,tmax,nt);
x=linspace(0,L,nx);
%Parameters
EI=2;rho=0.2;m=0.2;Ih=0.5;

dyLj=0;
yL_1=0.1;
yxx0_1=0;
yxxx_L0=0;
%Define viriables and Initial condition:
y=zeros(nx,nt);    %elastic deflectgion
th_2=0;th_1=0;
thd=0.5;dthd=0;ddthd=0;

l=0.50;
lamda10=0.55;     %>abs(e(0))=0.10
lamda1_inf=0.01;
lamda20=0.12;     %>abs(e(0))=0.10
lamda2_inf=0.01;

for i=1:1:nx
   y(i,:)=i^2/810;
end

for j=3:nt    %Begin
```

11.6 Simulation Example

```
        dth(j)=(th_1-th_2)/T;

        alfa1=10;alfa2=10;alfa3=10;alfa4=10;alfa5=10;alfa6=10;

        e1(j)=th_1-thd;
        de1(j)=dth(j)-dthd;

        yL(j)=yL_1;
        dyL(j)=dyLj;
        %%%%%%%%%%%%%%%%%%%%%%%%%%%%%%%

        lamda1(j)=(lamda10-lamda1_inf)*exp(-1*j*T)+lamda1_inf;
        S1(j)=e1(j)/lamda1(j);

        epc1(j)=0.5*log((1+S1(j))/(1-S1(j)));

        lamda2(j)=(lamda20-lamda2_inf)*exp(-1*j*T)+lamda2_inf;
        e2(j)=yL(j);
        de2(j)=dyL(j);
        S2(j)=e2(j)/lamda2(j);
        epc2(j)=0.5*log((1+S2(j))/(1-S2(j)));

        F(j)=alfa2*e2(j)-alfa4*epc2(j)*lamda2(j)/(lamda2(j)^2-e2(j)^2)-alfa6*
        de2(j);
        tol(j)=alfa1*e1(j)-alfa3*epc1(j)*lamda1(j)/(lamda1(j)^2-e1(j)^2)-alfa
        5*de1(j)-L*F(j);

        yxx0=(y(3,j-1)-2*y(2,j-1)+y(1,j-1))/dx^2;

        th(j)=2*th_1-th_2+T^2/Ih*(tol(j)+EI*yxx0);
        ddth(j)=(th(j)-2*th_1+th_2)/T^2;

        %get y(i,j),i=1,2, Boundary condition (3) or (9)
        y(1,:)=0;     %y(0,t)=0, i=1
        y(2,:)=0;     %y(1,t)=0, i=2

        %get y(i,j),i=3:nx-2
        for i=3:nx-2

        yxxxx=(y(i+2,j-1)-4*y(i+1,j-1)+6*y(i,j-1)-4*y(i-1,j-1)+y(i-2,j-1))/dx
        ^4;
```

```
        y(i,j)=T^2*(-i*dx*ddth(j)-EI/rho*yxxxx)+2*y(i,j-1)-y(i,j-2);    %i*dx=x
        ,(10)
    end

    %get y(nx-1,j),i=nx-1
    yxxxx(nx-1,j-1)=(-2*y(nx,j-1)+5*y(nx-1,j-1)-4*y(nx-2,j-1)+y(nx-3,j-1)
    )/dx^4;
    y(nx-1,j)=T^2*(-(nx-1)*dx*ddth(j)-EI/rho*yxxxx(nx-1,j-1))+2*y(nx-1,j-
    1)-y(nx-1,j-2);

    %get y(nx,j),y=nx
    yxxx_L=(-y(nx,j-1)+2*y(nx-1,j-1)-y(nx-2,j-1))/dx^3;
    y(nx,j)=T^2*(-L*ddth(j)+(EI*yxxx_L+F(j))/m)+2*y(nx,j-1)-y(nx,j-2);
    dyL(j)=(y(nx,j-1)-y(nx,j-2))/T;

    yL_1=y(nx,j);
    th_2=th_1;
    th_1=th(j);
    yxx0_1=yxx0;
    yxxx_L0=yxxx_L;
    dyLj=dyL(j);
    end    %End

    %To view the curve, short the points
    tshort=linspace(0,tmax,(nt-1)/100+1);
    yshort=zeros(nx-1,(nt-1)/100+1);
    for j=1:(nt-1)/100+1
        for i=1:nx
            yshort(i,j)=y(i,(j-1)*100+1);    %Using true y(i,j)
        end
    end
    figure(1);
    subplot(211);
    plot(t,thd*t./t,'r',t,th,'k','linewidth',2);
    legend('Ideal angle signal','Angle tracking');
    xlabel('Time');ylabel('Angle tracking');
    subplot(212);
    plot(t,dthd*t./t,'r',t,dth,'k','linewidth',2);
    legend('Ideal angle speed signal','Angle speed tracking');
    xlabel('Time');ylabel('Angle speed tracking');

    figure(2);
```

11.6 Simulation Example

```
surf(tshort,x,yshort);
colormap(summer);
title('Deflection of the flexible manipulator');
xlabel('time'); ylabel('x');zlabel('Deflection,y(x,t)');
colormap('Jet')

figure(3);
subplot(2,1,1)
plot(t,tol,'k','linewidth',2);
title('The control torque at the shoulder motor');
xlabel('Time');ylabel('Control input,tol');
subplot(2,1,2)
plot(t,F,'k','linewidth',2);
title('The control torque at the end actuator');
xlabel('Time');ylabel('Control input,F');

figure(4);
subplot(211);
for j=1:(nt-1)/100+1
    yshortL(j)=y(nx,(j-1)*100+1);
end
plot(tshort,yshortL,'k','linewidth',2);
xlabel('time');ylabel('y(L,t)');
subplot(212);
for j=1:(nt-1)/100+1
    yshort1(j)=y((nx-1)/2,(j-1)*100+1);
end
plot(tshort,yshort1,'k','linewidth',2);
xlabel('time');ylabel('y(x,t) at half of L');

figure(5);
subplot(211);
plot(t,e1,'k',t,-lamda1,'r--',t,lamda1,'r:','linewidth',2);
title('Position tracking error');
xlabel('Time');ylabel('Angle tracking error');
subplot(212);
plot(t,e2,'k',t,-lamda2,'r--',t,lamda2,'r:','linewidth',2);
title('Deflection at the end');
xlabel('Time');ylabel('Deflection at the end');

figure(6);
plot(t,dyL,'k','linewidth',2);
xlabel('time');ylabel('dy(L,t)');
```

References

1. C.P. Bechlioulis, G.A. Rovithakis, Adaptive control with guaranteed transient and steady state tracking error bounds for strict feedback systems. Automatica **45**(2), 532–538 (2009)
2. C.P. Bechlioulis, G.A. Rovithakis, Prescribed performance adaptive control for multi-input multi-output afine in the control nonlinear systems. IEEE Trans. Autom. Control **55**(5), 1220–1226 (2010)
3. C.P. Bechlioulis, G. Rovithakis, Robust adaptive control of feedback linearizable mimo nonlinear systems with prescribed performance. IEEE Trans. Autom. Control **53**(9), 2090–2099 (2008)
4. Z.J. Liu, J.K. Liu, Adaptive iterative learning boundary control of a flexible manipulator with guaranteed transient performance. Asian J. Control **19**(4), 1–12 (2017)
5. C.D. Rahn, *Mechatronic Control of Distributed Noise and Vibration* (Springer, New York, 2001)
6. G.H. Hardy, J.E. Littlewood, G. Polya, *Inequalities* (Cambridge University Press, Cambridge, 1959)
7. W.H. Ray, *Advanced Process Control* (McGraw-Hill, New York, 1981)
8. D. Christopher, *Rahn, Mechatronic Control of Distributed Noise and Vibration-A Lyapunov Approach* (Springer, Heidelberg, 2001)

Chapter 12
Conclusions

The book has been dedicated to PDE (partial differential equation) modeling and boundary controller design of flexible manipulator system. The results of the research work conducted in this book are summarized in each chapter, and the contributions made are reviewed. The key results are listed as follows.

In Chap. 3, to design PDE model and boundary conditions of flexible manipulator, we consider the flexible one-link manipulator that moves in the horizontal direction, the potential energy only depends on the flexural deflection of the link, Hamilton principle is used. Simulation results have demonstrated that the modeling is effective.

In Chap. 4, a composite boundary controller for flexible manipulator is presented based on the PDE model. The singular perturbation approach is designed, which results in two simple subsystems—slow subsystem and fast subsystem. Considering the characteristics of flexible manipulators, a composite controller for the full model is proposed which include an angle controller for the slow subsystem and a direct feedback controller for the fast subsystem to suppress the vibration. Simulation results demonstrate the effectiveness of the proposed controller.

In Chap. 5, we consider the motion and vibration of a flexible manipulator coupled and interacting with each other, which will interfere with the performance of the robot arm. To weaken the vibration of the flexible manipulator in motion, a boundary controller is designed with exponential convergence. Simulation results demonstrate the effectiveness of the proposed controller.

In Chap. 6, to overcome the shortcoming of Chap. 5, the boundary controller based on LaSalle analysis is designed, where $z_{xxx}(L)$ and $\dot{z}_{xxx}(L)$ are not needed in the controller design. In the closed system analysis, dissipative analysis, unique analysis and convergence analysis are given, the technique of separation of variables and extended LaSalle's invariance principle are used, asymptotic stability of the closed-loop system can be guaranteed. Simulation results are given to verify the effectiveness of the proposed controller.

In Chap. 7, the problem of full state constraints control is investigated for output constrained flexible manipulator system based on the PDE dynamic model. To prevent states from violating the constraints, a Barrier Lyapunov Function which grows to infinity whenever its arguments approach to some limits is employed. To regulate the joint angle and eliminate the elastic vibration, a boundary controller is developed. Furthermore, based on the Barrier Lyapunov Function and the boundary controller, we can guarantee that full state constraints and output tracking can be achieved simultaneously. The stability of the closed-loop system is carried out by the Lyapunov stability theory. Numerical simulations are given to illustrate the performance of the closed-loop system.

In Chap. 8, we consider the boundary control problem of the flexible manipulator in the presence of input saturation, a boundary control scheme is designed to regulate angular position and suppress elastic vibration simultaneously. The proposed control scheme allows the application of smooth hyperbolic functions, which satisfy physical conditions and input restrictions, easily be realized. It is proved that the proposed control scheme can be guaranteed in handling control input saturation. The stability is achieved through rigorous analysis without any simplification of the dynamics. Numerical simulations demonstrate the effectiveness of the proposed scheme.

In Chap. 9, we designed a robust observer for a flexible single-link manipulator based on the PDE dynamic model. Unlike the previously introduced observers for PDE model, this observer is designed to estimate the distributed spatiotemporally varying states with unknown boundary disturbance and spatially distributed disturbance. The asymptotic stability of the proposed observer is proved by theoretical analysis and demonstrated by simulation results.

In Chap. 10, we designed an infinite dimensional disturbance observer for flexible manipulator based on the PDE dynamic model. According to the designed inequality, we can conclude that the disturbance estimate errors are exponential convergence and the disturbance estimates converge to the true values exponentially. Numerical simulations demonstrate the effectiveness of the proposed observer.

In Chap. 11, considering a boundary control problem of a flexible manipulator with output constraints, we introduce a basic boundary controller design method with guaranteed transient performance. With the Lyapunov's direct method, a boundary controller is designed to regulate the angular position and suppress elastic vibration simultaneously. The proposed control scheme allows the errors to converge to an arbitrarily small residual set, with convergence rate larger than a pre-specified value. Numerical simulations demonstrate the effectiveness of the proposed scheme.

In summary, this book covers the dynamical analysis and control design for flexible manipulator system. The book is primarily intended for researchers and engineers in the control system. It can also serve as a complementary reading on modeling and control of mechanical systems at the post-graduate level.